Ice and Construction

Ice and Construction

**State-of-the-Art Report prepared by RILEM
Technical Committee TC-118, Ice and Construction**

RILEM
(The International Union of Testing and Research
Laboratories for Materials and Structures)

Edited by

L. Makkonen
Technical Research Centre of Finland, Espoo, Finland

CRC Press
Taylor & Francis Group
Boca Raton London New York

CRC Press is an imprint of the
Taylor & Francis Group, an **informa** business
A TAYLOR & FRANCIS BOOK

CRC Press
Taylor & Francis Group
6000 Broken Sound Parkway NW, Suite 300
Boca Raton, FL 33487-2742

First issued in paperback 2019

© 1994 RILEM
CRC Press is an imprint of Taylor & Francis Group, an Informa business

No claim to original U.S. Government works

ISBN-13: 978-0-419-20020-8 (hbk)
ISBN-13: 978-0-367-44928-5 (pbk)

A catalogue record for this book is available from the British Library

Publisher's Note
The publisher has gone to great lengths to ensure the quality of this reprint but points out that some imperfections in the original may be apparent

Visit the Taylor & Francis Web site at
http://www.taylorandfrancis.com

and the CRC Press Web site at
http://www.crcpress.com

Contents

Preface

This is a state-of-the-art report prepared by RILEM (International Union of Testing and Research Laboratories for Materials and Structures) Technical Committee TC-118. It reviews the use of ice and the problems caused by ice in construction. The problems reviewed cover a wide range of engineering applications and physical processes. Comprehensive reference lists are included and recommendations for future work are given. Two specific fields of problems, design ice forces on offshore structures and frost heaving, are outside the scope of this report, as they have been recently extensively reviewed in other connections.

RILEM TC-118 worked in the period 1988-1992 and had meetings in Canada, Finland and Sweden. In addition the Chairman had several meetings with the members of the Committee individually.

The principal authors of this report are:

Lasse Makkonen, Technical Research Centre of Finland (Chapters 1, 2.1, 2.4, 2.5, 4 and 6)
Wilfrid A. Nixon, University of Iowa, USA (Chapter 2.3)
Nirmal K. Sinha, National Research Council of Canada (Chapter 2.2)
Seppo Huovinen, Technical Research Centre of Finland (Chapter 5)
Eila Lehmus, Technical Research Centre of Finland (Chapters 3.1 and 3.2)
Lennart Fransson, Luleå University of Technology, Sweden (Chapter 3.3)

The members of TC-118 were:

Dr P. Duval, Laboratoire de Glaciologie/CNRS, France
Prof. L. Elfgren, Luleå University of Technology, Sweden
Dr R. Ettema, University of Iowa, USA
Dr B.C. Gerwick Jr, Ben C. Gerwick Inc., USA
Dr L. Makkonen, Technical Research Centre of Finland (Chairman)
Prof. W.M. Sackinger, University of Alaska, USA
Dr N.K. Sinha, National Research Council of Canada

In addition, the following persons, representing absent Members, have participated in the meetings: Dr Lennart Fransson, Dr Tuomo Kärnä and Mr Lars Stehn. Dr Åke Skarendahl has been the Counsellor of the Technical Committee.

As the chairman of RILEM TC-118 I wish to thank the above mentioned individuals and their supporting organizations for their contributions to the work of the Technical Committee and for preparing this report.

Lasse Makkonen
Espoo, August 1993

1

Introduction

Ice is a relatively strong material that, in cold regions, is abundant and inexpensive to manufacture. This gives it a high potential in construction, especially considering that transportation of other building materials to remote arctic areas is very expensive. The problem, of course, is that even in the coldest areas of the world ice is so close to its melting temperature that its mechanical properties are strongly temperature-dependent, and melt protection is usually necessary. The engineering properties of ice, the applications, the construction techniques and solutions of the special problems are reviewed in Chapters 2 and 3.

Ice is a serious problem for traditional construction in cold regions. For many structures in cold regions, such as those offshore, tall masts, and buildings on permafrost, ice forces are the major design criteria. The ice loads caused by sea ice on offshore structures have recently been thoroughly reviewed in two state-of-the-art reports (American Concrete Institute, 1985; Timco, 1989). Therefore, these matters are purposely set aside in this report. The same applies to building on permafrost, for which the relevant reviews are Johnston (1981) and Johnson (1986).

As for the testing methods of ice as a material, the reader is referred to a recent report (Häusler, 1988), and this subject is only briefly discussed here. Major engineering problems, such as avalanche protection, snow drifting and snow loads on roofs, are related to ice in the form of snow. These aspects are discussed by, for example, Schriever and Otstavnov (1967) and Gray and Male (1981), and are outside the scope of this report. All this leaves two major problem areas to be discussed here in detail: accretion of ice on structures in Chapter 4 and abrasion of concrete structures by ice in Chapter 5.

REFERENCES

American Concrete Institute (1985) *State-of-the-Art Report on Offshore Concrete Structures for the Arctic*. ACI Report 357. 1R-85, 120 pp.

Gray, D.M. and Male, D.H. (1981) *Handbook of Snow*, Pergamon Press, Canada, 776 pp.

Häusler, F.U. (ed.) (1988) IAHR Recommendations on testing methods in ice, 6th Report of the Working Group on Testing Methods in Ice. *9th International Association for Hydraulic Research (IAHR) Ice Symposium*, Sapporo, Japan, 24 pp.

Johnson, T.C., Berg, R.L., Chamberland, E.J. and Cole, D.M. (1986) *Frost Action Predictive Techniques for Roads and Airfields*. US Army Cold Regions Research and Engineering Laboratory Report 86-18, 45 pp.

Johnston, G.H. (ed.) (1981) *Permafrost Engineering, Design and Construction*, Wiley and Sons, 540 pp.

Schriever, W.R. and Otstavnov, V.A. (1967) *Snow Loads – Preparation of Standards for Snow Loads*. International Council for Building Research (CIB), Research Report 9, pp. 13–33.

Timco, G.W. (Editor) (1989) *International Association for Hydraulic Research (IAHR) Working Group on Ice Forces – 4th State-of-the-Art Report*. US Army Cold Regions Research and Engineering Laboratory Special Report 89-5, 385 pp.

2

Ice as a construction material

2.1 POSSIBLE APPLICATIONS

2.1.1 Permanent construction

Ice creeps (for a detailed discussion of the creep of ice, see Hobbs, 1974). Therefore, the word 'permanent' in connection with ice essentially means nonseasonal: that is, ice structures that last for years or decades. In time-scales longer than this even the polar ice-caps and glaciers are not truly permanent, but are results of a balance between ice outflow and snow accumulation.

Because of the creep and the fact that in almost all cold areas of the world the outside temperature occasionally exceeds the melting point of ice, unprotected ice structures have a limited lifetime. This can be much improved by melt protection (section 2.4): for example, by covering the structure with insulating materials. In the presence of such materials, however, adding new ice as needed becomes more difficult.

If the inside of an ice structure is kept at a normal room temperature, very effective insulation is also required inside the structure. For this reason, using ice as a material has more appealing applications in unheated buildings, such as those used for storage or vehicle shelters. Indeed, buildings where a cold inside temperature is also required in summer are among the most suitable applications of ice. Only a few attempts to use the cooling capacity of ice in construction have been made so far, but an example of a cold storage partly made of ice is shown in Figs 2.1 and 2.2. In this building, however, all supporting members are made of steel, and the purpose of the ice walls is to keep the inside air cool. New ice is added to the walls (by compacting snow) in winter. A similar method to make year-round winter sports facilities has been suggested recently (Makkonen *et al.*, 1993).

Fig. 2.1 Snow storage for summer use. City of Tokamachi, Japan.

Fig. 2.2 Interior view of the cold storage in Fig. 2.1.

There are no other man-made permanent nonrefrigerated structures known to the authors except for a military camp constructed in ice (entirely under the ice surface of a glacier) by the US Army in the 1960s. However, natural ice structures have been frequently utilized. Ice islands breaking off from ice shelves have been used as bases for scientific observations. These observation stations have been supported by aircraft landing on the ice islands. In a sense, wharves and airfields constructed on ice islands, and in Antarctica (Mellor, 1988; Mellor *et al.*, 1991), as well as insulated airfields on permafrost in Alaska (Crory, 1991), are also permanent ice constructions.

2.1.2 Artificially cooled structures

Although it is a structure in a limited sense, the skating rink is by far the most common permanent form of artificially cooled ice. Bobsled tracks are only slightly more complicated structures. However, many efforts have also been made to plan and build refrigerated permanent structures of considerable size and structural complexity.

The most prominent of these was the proposed Habbakuk project by the Allies in 1942–1944 (Gold, 1990). The aim was to use ice in building aircraft carriers with a length of 610 m and a draft of about 46 m. The plan was never completed, but a prototype vessel with a length of 18.3 m and a draft of 6 m was built (Gold, 1990). Later, quays (Marthinsen, 1986), dams (Renkel, 1991) and waterfronts (Faiko, 1990) have been built of ice and refrigerated. Artificially maintained permafrost has been used as a foundation for structures, such as oil pipelines and buildings.

The most promising applications of permanent refrigerated structures are in remote cold areas where other building materials are scarce and the need for refrigeration energy is small. In these areas it is also possible to use heat pipes in winter to provide cold-storage energy in the structure or foundation.

2.1.3 Seasonal construction using ice

Perhaps the oldest and best-known seasonal ice construction is the Eskimo igloo. But also in modern times, as in mining and exploration activities and in recreational areas for example, temporary enclosures are required for winter. Often the necessary lifetime of these enclosures is only a few months, in which case ice may well be used as the construction material regardless of the melting that takes place in spring.

Man-made ice may also be used as foundations for oil and gas exploration offshore. Massive ice structures can replace artificial gravel islands. The feasibility and economical benefits of these artificial ice islands have been demonstrated during the last few years (Cox and Utt, 1986; Goff and Masterson, 1986; Juvkam-Wold, 1986; Prodanovic, 1986). Another appli cation of a massive offshore ice structure is a barrier that protects an oil-drilling platform from the forces exerted by drifting sea ice (Kemp *et al.*, 1988).

Winter roads on both lake and sea ice, and snow roads on marshlands, can also be classified as seasonal construction, in which methods to accelerate the freezing process are often used. Even railroads have been built on seasonal ice (Kubo, 1988). These applications are discussed in more detail in Chapter 3.

Ice is a very suitable material for winter entertainment, such as ice and snow sculptures and other recreational structures (Lillberg, 1993). Very massive buildings can, and have been built. In a recent winter exhibition in Harbin, China, 22 000 tonnes of ice were used for the ice buildings.

While it is natural that seasonal ice buildings are for winter use, it should be mentioned that seasonal ice structures especially for summer use have also been suggested (Makkonen and Tuovinen, 1993; Makkonen *et al.*, 1993). These could be useful in applications where the cold-storage capacity of the structure can be used for cooling: for example, in storing foodstuff and medicine over the summer or in making winter sports facilities for summer use.

References

Cox, G.F.N. and Utt, M.E. (1986) Ice properties in a grounded man-made ice island, in *Proceedings of 5th International Conference on Offshore Mechanics and Arctic Engineering (OMAE)*, Tokyo, Japan, ASME, Vol. IV, pp. 135–42.

Crory, F.E. (1991) *Construction guidelines for oil and gas exploration in Northern Alaska*. US Army Cold Regions Research and Engineering Laboratory Report 91–21, 83 pp.

Faiko, L.I. (1990) Experience of construction large ice structures. *Acad. Nauk USSR, Data of Glaciological Studies*, Vol. 68, pp. 104–6 (in Russian).

Goff, R.D. and Masterson, D.M. (1986) Construction of a sprayed ice island for exploration, in *Proceedings of 5th International Conference on Offshore Mechanics and Arctic Engineering (OMAE)*, Tokyo, Japan, ASME, Vol. IV, pp. 105–110.

Gold, L.W. (1990) *The Canadian Habbakuk Project*, International Glaciological Society, 323 pp.

Hobbs, P.V. (1974) *Ice Physics*, Clarendon Press, Oxford, 837 pp.

Juvkam-Wold, H.C. (1986) Spray-ice islands evaluated for Arctic-building structures. *Oil and Gas Journal*, April, pp. 57–66.

Kemp, T.S., Foster, R.J. and Stevens, G.S. (1988) Construction and performance of the Kadluk 0-07 sprayed ice pad, in *Proceedings of 9th International Conference on Port and Ocean Engineering under Arctic Conditions (POAC)*, Fairbanks, Alaska, Vol. III, pp. 551–64.

Kubo, Y. (1988) Historical study of ice engineering. Unpublished presentation at the 9th International Association for Hydraulic Research (IAHR) Ice Symposium, Sapporo, Japan.

Lillberg, J. (1993) Snow and ice sculpting conquer the World. *IGS Symposium on Applied Ice and Snow Research*, Rovaniemi (abstract).

Makkonen, L. and Tuovinen, J. (1993) Buildings made of man-made ice. *IGS Symposium on Applied Ice and Snow Research*, Rovaniemi (abstract).

Makkonen, L., Hakala, R. and Yliniemi, P. (1993) *A cross-country skiing facility for year-around use*. Finnish patent No. FI-90372, 10 pp. (in Finnish)

Marthinsen, A. (1986) Ice used as a permanent construction material, in *Proceedings of 5th International Conference on Offshore Mechanics and Arctic Engineering (OMAE)*, Tokyo, Japan, ASME, Vol. IV, pp. 120–32.

Mellor, M. (1988) *Hard-surface runways in Antarctica*. US Army Cold Regions Research and Engineering Laboratory Special Report 88–13, 87 pp.

Mellor, M., Barthelemy, J.L., Fitzsimmons, G.J., Haehnle, R.J. and Weeks, W.F. (1991) *Ice wharf enquiry*. Division of Polar Programs, National Science Foundation, Report MP 2945, Washington D.C., 18 pp. + attachments.

Prodanovic, A. (1986) Man-made ice island performance, in *Proceedings of 5th International Conference on Offshore Mechanics and Arctic Engineering (OMAE)*, Tokyo, Japan, ASME, Vol. II, pp. 89–95.

Renkel, A.F. (1991) Ice and snow – construction materials. *Stroitel'stvo Truboprovodov*, **2**, 37–9 (in Russian).

2.2 STRENGTH OF ICE

2.2.1 The concept

Engineering designs are based on operational lifetime and a measure of the extreme values of the properties of the materials involved in the design. The strength of ice has played a central role in ice engineering. A knowledge of the local and the global pressure that ice can exert on a structure is essential for design purposes, as the simplest solution methods are based on the concept of strength properties of ice. Ice has been treated as a material at low temperatures and, like other geotechnical materials, the approaches and methodologies of traditional engineering schools were followed in developing the field of ice engineering. As a consequence, considerable effort was made in obtaining numerical values of the tensile and compressive strength and failure envelope for ice.

What is 'strength of ice'? Conventionally, the strength of a material is defined as the maximum load that a material body can support. This definition has also been adopted for ice. Both laboratory and field tests were conducted in the past for determining the ice strengths required in engineering calculations.

Leaving aside the fracture-mechanics tests, mechanical tests are primarily divided into two methods (Mellor and Cole, 1982, 1983): uniaxial or multi-axial constant strain or deformation rate (CD) tests and uniaxial constant stress (CS) deformation. Ideally, a displacement rate is suddenly imposed on a cylindrical or prismatic specimen and held constant in CD tests; in CS tests, a stress is suddenly imposed and held constant. CS tests are known as **creep tests**; CD tests are known as **strength tests**.

CD tests on ice are performed by making use of commercial testing machines. Usually the displacement rate of the cross-head or the actuator is maintained constant during a CD test. Stress is recorded against displacement, and stress–strain diagrams are drawn. The initial slope of the stress–strain diagram is used to calculate the elastic modulus, and the peak stress is used as the strength. In the tests, it is usually the load that is maintained constant. The usual practice is to record strain–time curves, from which instantaneous strain rate is determined as a function of time.

In the field of ice engineering, CD tests are preferred because these tests readily provide a means for determining both elastic moduli and strengths (Michel, 1978; Lainey and Tinawi, 1984). It should be pointed out here that making use of the stress–strain diagram for determining the modulus and the strength is based on the inherent assumption that the material response is independent of time. This implicit assumption had a profound influence on how ice was treated as a material. It had far-reaching effects, and engineers tend to abstain from conducting CS tests on ice. As a result, there are no uniaxial creep data on sea ice in the open literature. Yet the information is vital for analysing the bearing capacities of sea-ice covers or platforms such as spray-ice islands. The lack of creep data on sea ice provides an example of the failure of conventional wisdom. Actually, CS tests can be used for determining both the modulus and the strength of ice (Sinha *et al.*, 1992). In addition, the results are very useful for the development of constitutive equations. Moreover, CS tests can be performed using very cheap equipment.

Laboratory results are limited to small and selected volumes of materials. There is, therefore, a requirement for field tests and field equipment. A number of portable instruments have been developed in the last two decades for the determination of the *in-situ* strength of ice. A few theoretical frameworks have also been developed in order to interpret the field results. A dialogue, however, is necessary to clarify, at the outset, some apparent peculiarities of ice as a material and its strength properties.

2.2.2 Compression

The strength of ice is not a unique property. In reality, the strength numbers generated by CD tests are, at best, index values and are useful only for certain limited practical uses. Because ice is a natural material, its properties vary significantly, owing to the natural variations in the material. The structure, the texture and the fabric of the aggregate under consideration determine its mechanical properties and the mode of failure. Consequently, empirical equations have been developed for describing the strength as a function of density or porosity, caused by variations in air or brine volume.

What baffles engineers is the rate sensitivity and load-path dependence of the material properties. Both the strength of ice and the mode of failure depend on the loading rate and the history of loading. Even the slope of the stress–strain diagram, conventionally used for determining the elastic modulus, depends on the loading rate. There is no choice but to report strength and modulus in terms of strain rate or, more commonly, in terms of nominal strain rates, estimated from the cross-head rate and the specimen length. Yet the vital information on stiffness characteristics of test systems and histories of stress, time of failure, three-dimensional specimen strain, cracking activities, etc., necessary for quantitative analyses of such test results, are usually not presented.

The observed strain-rate sensitivity has, however, led to the questions: what strain rate is applicable for an indentation test, or what strain rate applies to an ice-structure interaction process? These questions, however, led to the absurd methods of geometrical normalization, such as the ratio of the velocity of an ice cover and some multiples of the dimension of the structure. When we focus our effort on the wrong target, in this case strain rate, we depart from our ability to solve a problem.

Depending upon the rate of loading, ice can behave as a brittle material or as a ductile material. A picture of a glacier makes one think that ice flows like butter. Drop an ice cube on the floor: it shatters like glass. Thus ice can be very ductile and it can also be very brittle. To alleviate this difficult problem and to apply the conventional wisdom, based on strength as a unique number, a simple concept of brittle/ductile transition has been introduced. Ice is assumed to behave as a brittle material in the 'brittle regime' above a certain rate of loading.

The ductile failure is usually referred to the failure mode in which the stress increases monotonically with strain to a maximum value and then decreases gradually with further deformation. This mode has been called an **upper-yield** type of failure by Sinha (1981). A strong relationship can be established between the strength in terms of either strain rate or stress rate:

$$\frac{\sigma_f}{\sigma_1} = A\left(\frac{\dot{\varepsilon}_{af}}{\dot{\varepsilon}_1}\right)^\alpha \tag{2.1}$$

where the subscripts f and 1 for stress indicate the maximum or failure value and the unit stress respectively; the subscripts af and 1 for strain rate indicate the average rate to the failure and the unit rate respectively.

The following experimental observations at $-10°C$ will give an idea of the response of different types of ice:

- $A = 212$ and $\alpha = 0.345$ for freshwater columnar-grained ice loaded normal to the axis of the columns (Sinha, 1981; 1982);

- $A = 45$ and $\alpha = 0.32$ for natural columnar-grained Arctic sea ice loaded normal to the axis of the columns (Sinha, 1984)
- $A = 139$ and $\alpha = 0.24$ for natural columnar-grained Arctic sea ice loaded parallel to the axis of the columns (Sinha, 1983a)
- $A = 71$ and $\alpha = 0.335$ for oriented, Arctic frazil sea ice (Sinha, 1986).

It should be noted that the value of the coefficient A depends significantly on the ice type and the loading direction, but the strain-rate sensitivity, given by the strain-rate exponent, does not vary with the type of ice. Similar observations were also noted when the strengths were examined on the basis of stress rate:

$$\frac{\sigma_f}{\sigma_1} = B \left(\frac{\sigma_{af}}{\sigma_1} \right)^{\beta} \tag{2.2}$$

In equation (2.2), the subscripts af and 1 for stress rate indicate the average rate to failure and the unit rate, respectively.

- $B = 11.7$ and $\beta = 0.30$ (Sinha, 1981a);
- $B = 12.4$ and $\beta = 0.23$ (Sinha, 1982) for freshwater columnar-grained ice loaded normal to the axis of the columns;
- $B = 4.64$ and $\beta = 0.29$ (Sinha, 1984) for natural columnar-grained Arctic sea ice loaded normal to the axis of the columns;
- $B = 24.9$ and $\beta = 0.34$ (Sinha, 1983a) for natural columnar-grained Arctic sea ice loaded parallel to the axis of the columns;
- $B = 6.7$ and $\beta = 0.30$ (Sinha, 1984) for oriented, Arctic frazil sea ice.

Again, it may be seen that the stress-rate sensitivity does not depend on the type of ice or loading conditions, although the coefficients vary significantly with ice type.

At high rates of loading, the uncertainty of loading conditions and the end effects have a profound influence on the outcome of the experimental results. The so-called **brittle–ductile transition** or the onset of premature failure has been found to depend on the microstructure, the state of the material and, most of all, on the test conditions and loading history. As the temperature increases, ice creeps more and its ductility increases (Barnes *et al.*, 1971). However, as the temperature increases, ice also becomes more brittle and cracks at lower stresses (Gold, 1972). The increase in ductility with increase in brittleness and increase in temperature has also been noted in sea ice (Sinha *et al.*, 1992). Ice seems to defy conventional wisdom and, as a consequence, this makes it even more difficult to grasp the complexities of ice failure. To an uninitiated engineer, ice, certainly, appears to be a strange material.

In addition to the multitude of structural features, natural ice includes many trapped impurities. The purity of ice depends on the purity of the water or the melt from which the ice grew. Freshwater ice contains impurities, usually in the form of air bubbles. These inclusions may pose little problem if blocks of freshwater ice are to be recovered and stored for tests to be conducted at later dates. Sea ice, however, may contain a significant amount of brine in the form of pockets trapped between the grains and the subgrains. Ideally, sea ice should be sampled, stored and transported at temperatures lower than -23 °C, the precipitation point of common salt crystals, $NaCl.2H_2O$. There is a marked tendency for brine loss if the ice is sampled when the ambient air temperatures are high, particularly higher than about -10 °C (Cox and Weeks, 1986).

2.2.3 *In-situ* field tests

Overcoming the acute problems of appropriate sampling and testing of sea ice is not simple. However, effort has been made in developing devices for *in-situ* tests. These instruments can be divided into three general groups: the **borehole pressuremeter** (Murat *et al.*, 1986; Shields *et al.*, 1989), the **borehole jack or indentor,** and the **flatjack** (Kivisild, 1975; Sinha, *et al.*, 1986; Sinha, 1987; Prowse, *et al.*, 1988; Iyer and Masterson, 1991). The National Research Council of Canada (NRCC) borehole indentor system was developed after evaluating existing test equipment and methodologies used in developing test techniques. It was designed and the test procedures were developed by keeping in mind that, from the physical point of view, ice is a high-temperature material.

Equation (2.2) can also be applied to *in-situ* borehole indentor tests (Sinha *et al.*, 1986). It was found that $B = 3.4$ and $\beta = 0.21$ at -10 °C, $B = 6.6$ and $\beta = 0.26$ at -20 °C, and $B = 9.1$ and $\beta = 0.24$ at -30 °C (Sinha *et al.*, 1986) for Arctic man-made built-up sea ice. It can be seen that β does not change with temperature in any systematic manner.

2.2.4 Tension

Polycrystalline ice is known to be significantly weaker in tension than in compression. Problems involving buckling or bending (or some complex loading conditions) tend to put an ice cover under tension. An ice body can fail essentially in one of two ways: propagation of an existing crack in the ice under the applied load, or initiation of a crack followed by propagation of the initiated crack. In the former case, the load should be sufficiently high to propagate the crack, and hence the mechanics of fracture play the dominant role in the failure processes of ice. In the latter case, cracks could be initiated inside the body of the ice by a number of processes, collectively known as **high-homologous-temperature micromechanisms**, as ice is so close to its melting temperature.

It is easy to perform beam-bending tests and obtain numbers for rupture strengths. These tests are popular, and a number of studies have been carried out in the past. However, these index values are difficult to interpret and to use primarily because of the nonlinear deformation response of ice.

Results for pure tensile tests are relatively easy to understand, but the tests are very difficult to perform. Tensile tests in polycrystalline metals, alloys and ceramics are normally performed using dumb-bell or dog-bone-shaped specimens. This type of geometry helps to generate the maximum stress and strain field in the gauge section and at the same time maintains the uniformity of the deformation field in this zone. The dumb-bell type of geometry has been used by Dykins (1967, 1969, 1970), Hawkes and Mellor (1972), and Burdick (1975). The dog-bone type of specimen has been used by Sinha (1989). The size of the specimen to be used is usually dictated by the available test facilities and the machinability of the material. As the grain sizes are small it is often assumed, and rightly so, that there are a large number of grains across the diameter or the minimum dimension of the gauge section.

The grain size of natural polycrystalline ice is usually several orders of magnitude larger than that of metals, alloys and ceramics. The specimen size for ice should therefore be huge compared with that for other materials if a sufficient number of grains (more than ten grains, according to Jones and Chew, 1981) is to be accommodated within the specimen to avoid any geometric effect. Providing an adequate specimen size for direct tensile tests on fine-grained, granular ice is not difficult (Hawkes and Mellor, 1972; Burdick, 1975) but it has been an acute problem in testing natural ice (Peyton, 1966). This is because of the difficulty of transporting large amounts of ice from the field to the laboratory. However, many tests on laboratory-grown saline ice have also been carried out in which this constraint was overlooked or ignored. Thin-section photographs of fractured specimens show only two grains (Fig. 30 in Dykins, 1970) and three grains (Fig. 15 in Dykins, 1967 and Fig. 6 in Dykins, 1969) across the specimen diameter.

All the authors mentioned above used specimens with reduced-section diameter or width in the range of 30 mm. Moreover, no direct measurements of strain in the gauge section were carried out. An improved method for tensile testing of ice and frozen soil, using special end caps and a pair of displacement gauges mounted on the gauge section, was presented by Eckardt (1982). This procedure was essentially followed by Kuehn et al. (1988), who used the Synthane end-caps described by Lee (1986).

Sea ice has commonly been seen to be columnar-grained, exhibiting either transverse isotropy or orthotropy in its crystallographic structure. For this type of ice, and for loads applied in the horizontal plane (the plane parallel to the surface of the ice cover, normal to the longitudinal axis of the grains), prismatic, rather than cylindrical, specimens are compatible with the grain structure. This is because a cylindrical specimen has nonuniform grain constraint for columnar-grained ice; the grains are of different lengths within the specimen, and the grain-boundary sliding mechanism plays an important role in the deformation and fracture processes. Moreover, a significant amount of valuable material (in the case of field samples) is wasted in making cylindrical specimens.

These considerations led Sinha (1978) to use prismatic specimens for creep as well as for strength tests on freshwater ice (Sinha, 1981) and sea ice (Sinha, 1984). An extension of the principle of using prismatic specimens for the tensile test is to use dog-bone-type specimens (Sinha, 1989). This technique has been adopted for testing first-year as well as multi-year sea ice. The method has been successfully applied in tensile testing of columnar-grained ice from a multi-year floe in the Canadian High Arctic.

Examination of the ice literature shows that investigators have in general ignored the rate sensitivity of the tensile strength of ice. The author has come across only one reference (Kubo, 1941) that proposed a rate sensitivity of the strength of freshwater ice as:

$$\sigma_f = \frac{\sigma_f}{6 \times 10 + 0.7\sigma_f} \qquad (2.3)$$

Examination of the available experimental results indicates that the tensile strength does show a small but definite rate sensitivity (Sinha, 1983).

References

Barnes, P., Tabor, D. and Walker, J.C.F. (1971) The friction and creep of polycrystalline ice. *Proceedings of the Royal Society*, **A324**, 127–55.

Burdick, J.L. (1975) Tensile creep-rupture of polycrystalline ice, in *Proceedings of 3rd International Conference on Port and Ocean Engineering under Arctic Conditions (POAC)*, Fairbanks, Alaska, University of Alaska, 11–15 August 1975, pp. 235–46.

Cox, G.F.N. and Weeks, W.F. (1986) Changes in the salinity and porosity of sea-ice samples during shipping and storage. *Journal of Glaciology*, **32** (112), 371–5.

Dykins, J.E. (1967) Tensile properties of sea ice grown in a confined system, in *Physics of Snow and Ice, Proceedings of International Conference on Low Temperature Science*, Sapporo, Japan (ed. H. Oura), Institute of Low Temperature Science, Hokkaido University, Vol. 1, Part 1, pp. 523–37.

Dykins, J.E. (1969) Tensile and flexture properties of saline ice, in *Physics of Ice, Proceedings of International Symposium on Physics of Ice*, Munich, West Germany (eds N. Riehl, B. Bullemer and H. Engelhardt), Plenum Press, New York, pp. 251–70.

Dykins, J.E. (1970) *Ice Engineering – Tensile Properties of Sea Ice Grown in a Confined System*. Technical Report R689, Naval Civil Engineering Laboratory, Port Hueneme, California.

Eckardt, H. (1982) Creep tests with frozen soils under tension and uniaxial compression, in *Proceedings of 4th Canadian Permafrost Conference, Calgary, Canada: The Roger J.E. Brown Memorial Volume* (ed. H.M. French), Associate Committee on Geotechnical Research, National Research Council of Canada, Ottawa, pp. 394–405.

Gold, L.W. (1972) The failure process in columnar-grained ice. *NRC Report TP 369*, Ottawa, 108 pp.

Hawkes, I. and Mellor, M. (1972) Deformation and fracture of ice under uniaxial stress. *Journal of Glaciology*, **11** (61), 103–31.

Iyer, S.H. and Masterson, D.M. (1991) Field strength values of multi-year ice off Herschel Island, in *Proceedings of 10th International Conference on Offshore Mechanics and Arctic Engineering (OMAE/ASME)*, Stavanger, Norway, 23–29 June 1991, Vol. 4, pp. 63 –70.

Jones, S.J. and Chew, H.A.M. (1981) On the grain-size dependence of secondary creep. *Journal of Glaciology*, **27** (97), 517–18.

Kivisild, H.R. (1975) Ice mechanics, in *Proceedings of 3rd International Conference on Port and Ocean Engineering under Arctic Conditions (POAC)*, University of Alaska, Vol. 1, pp. 287–313.

Kubo, Y. (1941) Construction Bureau, South Manchurian Railway Company Study on River Ice (in Japanese); private communication. See also US Army, Cold Regions Research and Engineering Laboratory, Draft Translation T50 (1955).

Kuehn, G.A., Lee, R.W., Nixon, W.A. and Schulson, E.M. (1988) The structure and tensile behavior of first year sea ice and laboratory-grown saline ice, in *Proceedings of 7th International Conference on Offshore Mechanics and Arctic Engineering (OMAE)*, 7–12 February, Houston, Texas, ASME, New York, Vol. 4, pp. 11–17.

Lainey, L. and Tinawi, R. (1984) The mechanical properties of sea ice – a compilation of available data. *Canadian Journal of Civil Engineering*, **11** (4), 884–923.

Lee, R.W. (1986) A procedure for testing cored ice under uniaxial tension. *Journal of Glaciology*, **32** (112), 540–1.

Mellor, M. and Cole, D.M. (1982) Deformation and failure of ice under constant stress or constant strain-rate. *Cold Regions Science and Technology*, **5**, 201–19.

Mellor, M. and Cole, D.M. (1983) Stress/strain/time relations for ice under uniaxial compression. *Cold Regions Science and Technology*, **6**, 207–30.

Michel, B.M. (1978) *Ice Mechanics*, Les Presses de l'Universite Laval, Quebec City, Canada.

Murat, J.R., Huneault, P. and Ladanyi, B. (1986) Effects of stress redistribution on creep parameters determined by a borehole dilatometer test, in *Proceedings of 5th International Conference on Offshore Mechanics and Arctic Engineering (OMAE/ASME)*, Tokyo, April 1986, ASME, Vol. 4, pp. 58–64.

Peyton, H.R. (1966) *Sea Ice Strength*. Geophysical Institute, University of Alaska, Fairbanks, Alaska, Report No. UAG R-182.

Prowse, T.D., Demuth, M.N. and Onclin, C.R. (1988) Using the borehole jack to determine changes in river ice strength, in *Proceedings of 5th Workshop on Hydraulics of River Ice/Ice Jams*, Winnipeg, Canada, June 1988, pp. 283–301.

Shields, D.H., Domaschuk, L., Funegard, E., Kjartanson, B.H. and Azizi, F. (1989) Comparing the creep behaviour of spray ice and polycrystalline freshwater ice, in *Proceedings of 8th International Conference on Offshore Mechanics and Arctic Engineering (OMAE/ASME)*, Houston, Texas, 19–23 March 1989, Vol. 4, pp. 235–45.

Sinha, N.K. (1981a) Rate sensitivity of compressive strength of columnar-grained ice. *Experimental Mechanics*, **21** (6), 209–18.

Sinha, N.K. (1981) Comparative study of ice strength data, in *Proceedings of International Association of Hydraulic Research (IAHR) International Symposium on Ice*, Quebec, Canada, Vol. 2, pp. 581–95.

Sinha, N.K. (1982) Constant strain- and stress-rate compressive strength of columnar-grained ice. *Journal of Materials Science*, **17** (3), 785–802.

Sinha, N.K. (1983a) Field tests on rate sensitivity of vertical strength and deformation of first-year columnar-grained sea ice, in *Proceedings, VTT Symposium 27, POAC '83*, Vol. 1, 1983, pp. 231–42.

Sinha, N.K. (1983b) Does the strength of ice depend on grain size at high temperatures? *Scripta Metallurgica*, **17** (11), 1269–73; also *Scripta Metallurgica*, **18** (12), pp. 1441–2.

Sinha, N.K. (1984) Uniaxial compressive strength of first-year and multi-year sea ice. *Canadian Journal of Civil Engineering*, **11** (1), 82–91.

Sinha, N.K. (1986) Young Arctic frazil sea ice: Field and laboratory strength tests. *Journal of Materials Science*, **21** (5), 1533–46.

Sinha, N.K. (1987) The borehole jack – is it a useful Arctic tool? *Journal of Offshore Mechanics and Arctic Engineering, Transactions of ASME*, **109** (4), 391–7.

Sinha, N.K. (1989) Ice and steel – a comparison of creep and failure, in *Mechanics and Creep of Brittle Materials* (eds A.C.F. Cocks and A.R.S. Pouter), Elsevier, London, pp. 201–12.

Sinha, N.K., Strandberg, A. and Vij, K.K. (1986) In situ assessment of drilling platform sea ice strength using a borehole jack, in

Proceedings of In Situ Testing and Field Behaviour, 39th Canadian Geotechnical Conference, 27–29 August 1986, Ottawa, Ontario, Canada, Canadian Geotechnical Society, pp. 153–7.

Sinha, N.K., Zhan, C. and Evgin, E. (1992) Creep of sea ice, in *Proceedings of 11th International Conference on Offshore Mechanics and Arctic Engineering (OMAE/ASME)*, Calgary, Alberta, 7–11 June 1992, Vol. 4.

2.3 REINFORCED ICE AND ADDITIVES

2.3.1 Rationale for reinforcement

The purpose of reinforcing any material is to provide an enhancement of various mechanical properties for that material. This holds true for ice, which exhibits a number of drawbacks when considered as a structural material. It is relatively weak, and extremely brittle. Furthermore, owing to the very high temperatures (relative to its melting point) that it typically experiences, ice usually exhibits significant creep deformation over time. Yet ice is used in a number of situations in northern climates as a building material. In general, it is used for temporary structures that will not survive more than one winter season (during the spring thaw the ice structure becomes unsafe and can no longer be used).

Ice reinforcement has been proposed as a way to improve the structural characteristics of ice. There are a number of benefits to be gained by reinforcing ice. The first is that the reinforcement will improve the tensile strength of ice, which is very low. The second is that the reinforcement of the ice will improve the time-dependent behaviour of the ice, by slowing down the creep rate. Both these improvements affect the mechanical behaviour of the ice directly. A third benefit that can be gained from reinforcement of ice is an improvement in the manufacturing time of an ice structure. If a given volume of material is to be produced by freezing, then that volume will be more rapidly frozen if part of the volume is taken up by nonfreezing materials (that is, solid materials). Further, by virtue of its strengthening effect, the reinforcement may mean that less total volume is required. In this way, adding reinforcement may significantly speed up the construction process, and thus provide considerable financial savings. The drawback is that the volume of material may disintegrate somewhat more rapidly during the spring thaw, owing to its reduced thermal capacity.

While reinforcing ice has benefits, it also has drawbacks. First, adding a second material to the construction process will increase both the difficulty and the cost of that process. If the reinforcement must be placed accurately there may be a significant cost increase in this regard, and a need for skilled labour. Further, if the reinforcing material is not available on site, shipping costs may be high. Unfortunately, many of the reinforcing materials described herein are not readily available in remote Arctic (or near Arctic) locations. Finally, care must be taken to ensure that the reinforcing material does not become an environmental hazard after thaw has occurred. This may require a costly clean-up process after the construction is no longer usable. All these factors may limit the use of reinforcement to a few, relatively small, locations, perhaps as part of a much larger structure in which particular strength is required. Under such limited conditions though, reinforced ice may well be the material of choice.

2.3.2 Types of reinforcement

Improvement of the tensile strength of ice can be achieved in several ways. First, other materials can be embedded in the ice in regions where it will experience tension. This material then has many similarities to reinforced concrete, in that the reinforcement is deliberately placed, and has a much higher tensile strength than the matrix in which it is embedded. The idea is that the reinforcement carries all or most of the tensile stress. In ice such reinforcement can be termed **macroscopic**, and has been achieved by use of tree trunks, steel, and geogrid.

The second way to improve the tensile strength of ice is to mix with the ice something that will inhibit crack formation and propagation, which tend in ice to determine the tensile strength. This sort of reinforcement is mixed homogeneously with the ice, to produce a material that is essentially isotropic. This differs from the macroscopic reinforcement described above, in which reinforcement is localized. In this case, it is generalized, and, in general, the scale of the reinforcing agent is similar to the microstructural scale of the ice. This homogeneous reinforcement can be termed **microscopic**. A number of reinforcing materials have been proposed for microscopic reinforcement of ice, including sawdust, asbestos, wood pulp, silt, sand and gravel.

(a) Macroscopic reinforcement
A number of studies have been conducted to examine the effects of

macro-reinforcement on the bearing strength of ice, and this sort of reinforcement has been used operationally (see Michel *et al.*, 1973) in the James Bay project to provide strengthened ice bridge crossings for trucks. Den Hartog (1975) discusses the use of various types of reinforcement (birch, trees and grass) and reports that reductions in required ice thicknesses of up to 25% can be achieved by use of such reinforcement. Den Hartog *et al.* (1976) report a series of tests of ice bridges made with trees for reinforcement, and indicate the load at which these full-scale prototype bridges failed. However, owing to difficulties during the freezing process, it was not possible to place the reinforcing trees (spruce was used) in the lower part of the ice sheet. Thus the reinforcement did not provide the benefit that might be expected from it.

This raises an important point for the use of macro-reinforcement. In general, when ice is used to support loads in bridges or roads, the maximum tensile stress is found on the bottom surface of the ice. Accordingly, this is where the reinforcement should ideally be placed, which is an extremely difficult task.

A number of studies have been conducted in both the field and the labora-tory on macro-reinforced ice. Jarret and Biggar (1980) performed laboratory tests on beams of ice reinforced with geogrid, and Haynes and Martinson (1989) tested the strength of ice sheets reinforced with geogrid. They placed the sheets of geogrid below the neutral plane of the sheet by cutting a slot into the ice sheet and feeding the geogrid through this slot. The geogrid was then held in place while the sheet froze down into and beyond it. An alternative method involved anchoring the geogrid to the bottom of the tank until a required thickness of ice had formed, then releasing the geogrid so that it floated up into contact with the bottom of the ice sheet. While both these methods ensured placement of the geogrid in the optimum location, it is not clear how effective they would be in the field.

A number of studies have been performed by Fransson and co-workers, using steel and wood bars as reinforcement, with these generally being placed in the top part of the ice sheet. Even though the reinforcement was not optimally placed, Fransson and Elfgren (1986) noted that sawn wood (and sand) provided good reinforcement, while birch branches provided no discernible reinforcing effect. A number of field tests were conducted and reported by Fransson (1983, 1985), in which cut wood and steel bars were used as reinforcement, and an increased ice cover strength was found as a result of the reinforcement. This confirmed earlier results reported by Cederwall and Fransson (1979) and Cederwall (1981).

Other laboratory studies by Grabe (1986) and Vasiliev (1986; personal communication, 1991) have investigated the possibility of using fibreglass rods and other such materials as reinforcement in ice. Again, as expected, the reinforcement increased the strength of the test samples. Stanley and Glockner (1976) and Glockner (1988) tested ice reinforced with fibreglass yarn, and proposed the use of this material for making reinforced ice domes in Arctic regions (for use as temporary shelters). Reinforcing domes in this way does increase the dome strength, especially if the reinforcement is placed at points of stress concentration in the dome (for example, around the door or other openings).

From the foregoing studies, it is clear that ice can be effectively reinforced by means of steel rods, wood dowels or cut lumber, and by fibreglass rods. However, it should be stressed that at present there is nothing even remotely resembling a design guide for the use of such materials in strengthening an ice cover. Accordingly, considerable care and experimental work would be needed before such reinforcement could be employed in actual projects. Nonetheless, the evidence of the James Bay project strongly suggests the usefulness of reinforcing ice in such a manner.

There are two key problems that must be addressed. First, can the reinforcement be placed so as to bear the tensile stresses within the ice? In general, this requires placement of the reinforcement under the ice sheet, which is difficult at best. However, even if the reinforcement cannot be placed optimally, benefit may still be gained from it, provided it can sustain compressive loading effectively. Fransson (1985) warns that some materials (he notes birch branches in particular) are not effective in this regard. The second problem facing the designer is how much and what type of reinforcement to use. Some insight may be gained into this using the basic theory behind reinforced concrete design, but until a more complete design guide is available, experiments, certainly in the laboratory and ideally at prototype scale in the field, will be needed to determine the optimum reinforcement to be used.

(b) Microscopic reinforcement

The first reported attempt to use ice with microscopic reinforcement occurred during the Second World War, as part of Operation Habbakuk. This was a plan, devised with Churchill's approval, to build a fleet of iceberg aircraft carriers to provide fighter cover for the North Atlantic convoys. To achieve a suitable resistance to torpedo damage for these iceberg aircraft carriers, a reinforced ice was proposed. This was a mixture of ice and sawdust, and was christened 'Pykecrete' after the person (Dr Geoffrey Pyke) who had proposed the iceberg aircraft carrier scheme. Perutz (1947) reports some of the strength test results obtained in the development of this rather fantastic project, and a more complete description is presented by Gold (1990). Perutz (1947) reports that adding between 5 and 10% sawdust to the ice increased bending strength by factors of five or greater.

Since the Second World War, a number of studies on the reinforcement of ice have been performed. Coble and Kingery (1962) reinforced ice with a number of different materials, and found that the largest increase in strength was from (in decreasing order) fibreglass, asbestos, newspaper mash, bond paper, wood pulp, starch, and bond paper strips. Abele (1964) added sawdust to mixtures of snow and ice, and obtained substantial strengthening. Nixon and Smith (1987) measured the fracture toughness of ice reinforced with a variety of wood-based materials (newspaper, wood pulp, sawdust, blotting paper and bark).

The fracture toughness is the inherent resistance of a material to fracture and crack propagation. Nixon and Smith (1987) proposed a strengthening model, which indicated that the fibrous wood-based material could provide strengthening in three ways. Each of these ways increased the energy required to fracture the ice. Thus, for an ice composite in which the reinforcing agent is fibrous, for fracture to occur the fibres must debond from the ice, they must be pulled out of the ice, and finally the fibres must fracture. Experimental results indicated that the fracture toughness was proportional to the inverse square root of the fibre diameter, and this suggests that the major reinforcing mechanisms are debonding and pull-out of the fibres. The extent to which these results are applicable to non-wood-based materials is unclear.

Kuehn and Nixon (1988) reported a number of experiments reinforcing both freshwater and saline ice with wood-based materials. The saline ice samples were very weak compared with the freshwater ice samples, largely because the sample-manufacturing method did not allow for brine drainage. Again, substantial increases in both fracture toughness and bending strength were realized when reinforcement was used. The authors also performed a simple cost analysis for using reinforced ice and, as might be expected, the use of reinforced ice indicated significant savings, though no site clean-up costs were considered.

More recently, Nixon and Weber have been considering the possibility of using silts, sands and gravels as the reinforcing agents in ice (Nixon, 1989; Nixon and Weber, 1990, 1991; Weber, 1990; Weber and Nixon, 1990, 1991). They performed three-point bend tests on ice beams reinforced with four different soils, with a range of reinforcement from 1.4 to 66.9% soil by volume, at three temperatures (−5, −10, −20 °C), at extreme fibre strain rates between 0.01 and 0.00001 s^{-1}, and using four beam sizes (spans of 305, 762, 1524 and 2135 mm). Soil has somewhat of an advantage over wood-based products in that it is likely to be available even in the most remote Arctic locations including offshore, and thus materials transportation costs can be considerably reduced by its use. Nixon and Weber found that the strength of the soil–ice composite beams was a clear function of the mean soil diameter (again, there was an inverse square root dependence), and developed a model that included all the variables mentioned above. While this model is semi-empirical, and requires much further elucidation, it does provide a starting point for a more complete description of ice reinforced by soils.

From the above it can be seen that micro-reinforcement of ice with a variety of products can lead to a material that is significantly stronger than plain ice. Some of the materials used in the past would not even be considered today (asbestos fibres for example), and one of the concerns in micro-reinforcement of ice is that the material can be collected after use, and safely removed so as to minimize any long-term environmental impact. This would not be required were the reinforced ice to be used as a permanent construction material (Marthinsen, 1986), but such applications do not seem currently feasible. Nevertheless, significant steps have been made towards the development of an understanding of how various materials can reinforce ice, and while the field is at present a considerable way away from implementing any sort of design guide, guidelines for ice reinforcement could begin to be developed over the next few years.

References

Abele, G. (1964) *Some Properties of Sawdust–Snow–Ice Mixtures.* US Army Cold Regions Research and Engineering Laboratory Special Report SR-60.

Cederwall, K. and Fransson, L. (1979) *The Effect of Reinforcement on the Carrying Capacity of an Ice Cover*, Publication 79-1, Division of Structural Engineering, University of Luleå (in Swedish).

Cederwall, K. (1981) Behavior of reinforced ice cover with respect to creep, in *Proceedings of 6th International Conference on Port and Ocean Engineering Under Arctic Conditions (POAC)*, University of Laval, Quebec, Vol. 1, pp. 562–70.

Coble, R.L. and Kingery, W.D. (1962) Ice reinforcement, in *Proceedings of Ice and Snow* (ed. W.D. Kingery), MIT Press, pp. 130–48.

Den Hartog, S.L. (1975) Floating ice for crossings, *The Military Engineer*, **67** (436), 64–6.

Den Hartog, S.L., McFadden, T. and Crook, L. (1976) *Failure of an Ice Bridge*. US Army Cold Regions Research and Engineering Laboratory Report 76-29.

Fransson, L. (1983) Full scale tests of the bearing capacity of an ice sheet, in *Proceedings of 7th International Conference on Port and Ocean Engineering Under Arctic Conditions (POAC)*, Helsinki, Finland, Vol. 2, pp. 687–97.

Fransson, L. (1985) Load bearing capacity of an ice cover subjected to concentrated loads, in *Proceedings of 4th International Conference on Offshore Mechanics and Arctic Engineering (OMAE)*, Dallas, Texas, Vol. II, pp. 170–6.

Fransson, L. and Elfgren, L. (1986) Field investigation of load–curvature characteristics of reinforced ice, in *Proceedings of POLARTECH 86*, Vol. 1, pp. 175–196.

Glockner, P.G. (1988) Reinforced ice domes as temporary enclosures for cold regions, in *Proceedings of 7th International Conference on Offshore Mechanics and Arctic Engineering (OMAE)*, Houston, Texas, Vol. IV, pp. 185–92.

Gold, L.W. (1990) *The Canadian Habbakuk Project*, International Glaciological Society, 323 pp.

Grabe, G. (1986) Reinforced ice as a construction material – creep of reinforced ice, in *Proceedings of POLARTECH 86*, Vol. 2, pp. 793–806.

Haynes, F.D. and Martinson, C.R. (1989) Ice reinforced with geogrid, in *Proceedings of 8th International Conference on Offshore Mechanics and Arctic Engineering (OMAE)*, The Hague, Netherlands, ASME, Vol. IV, pp. 179–85.

Jarret, P.M. and Biggar, K.W. (1980) *Ice Reinforcement with Geotechnical Fabrics*, NRC Canada, Association Comm. Geotechnical Resources, Technical Memo No. 129, pp. 60–68.

Kuehn, G.A. and Nixon, W.A. (1988) Reinforced ice: mechanical properties and cost analysis for its use in platforms and roads, in *Proceedings of 7th International Conference on Offshore Mechanics and Arctic Engineering (OMAE)*, Houston, Texas, Vol. IV, pp. 193–200.

Marthinsen, A. (1986) Ice used as a permanent construction material, in *Proceedings of 6th International Conference on Offshore Mechanics and Arctic Engineering (OMAE)*, Houston, Texas, Vol. IV, pp. 120–8.

Michel, B., Drouin, M., Lefebvre, L.M., Rosenberg, P. and Murray, R. (1973) Ice bridges of the James Bay Project. *Canadian Geotechnical Journal*, **11** (4), 599–619.

Nixon, W.A. (1989) Alluvium reinforced ice: a preliminary report of bending strength tests. *Cold Regions Science and Technology*, **16**, 309–13.

Nixon, W.A. and Smith, R.A. (1987) The fracture toughness of some wood–ice composites. *Cold Regions Science and Technology*, **14**, 139–45.

Nixon, W.A. and Weber, L.J. (1990) The effect of specimen size on the bending strength of alluvium reinforced ice, in *Proceedings of 9th International Conference on Offshore Mechanics and Arctic Engineering (OMAE)*, Houston, Texas, ASME, Vol. IV, pp. 217–22.

Nixon, W.A. and Weber, L.J. (1991) Flexural strength of sand reinforced ice. *Journal of Cold Regions Engineering, ASCE*, **5** (1), 14–27.

Perutz, M.F. (1947) A description of the iceberg aircraft carrier, *Journal of Glaciology*, **1** (3), 95–104.

Stanley, R.G. and Glockner, P.G. (1975) Reinforced ice, in *Proceedings of 3rd International Conference on Port and Ocean Engineering Under Arctic Conditions (POAC)*, Fairbanks, ALaska, Vol. 2, pp. 935–56.

Vasiliev, N.K. (1986) *Reinforcement of ice with dispersive and fiberglass materials* (in Russian). Vsesoiuznyi Naucho-issledovatel'skii Institute Gdrotekhniki, Leningrad, Izvestia, Vol. 188, pp. 54–8.

Weber, L.J. (1990) *A Study of the Flexural Properties of Alluvium Reinforced Ice Beams*, M.S. Thesis, Iowa Institute of Hydraulic Research, University of Iowa, Iowa City, IA 52242–1585.

Weber, L.J. and Nixon, W.A. (1990) The effect of soil type on the bending strength of alluvium reinforced ice, in *Proceedings of 10th*

International Association for Hydraulic Research (IAHR) Ice Symposium, Helsinki, Finland, Vol. 1, pp. 486–99.

Weber, L.J. and Nixon, W.A. (1991) The effect of loading rate on the bending strength of alluvium reinforced ice, in *Proceedings, ASCE Cold Regions Engineering Specialty Conference*, pp. 71–84.

2.4 MELTING AND PROTECTION OF ICE STRUCTURES

Ice structures may deteriorate by fracture, creep and melting. Protection against fracture and creep must be mostly done in the construction phase by reinforcement, as discussed in section 2.3. Here we discuss thermal protection against melting. To some extent thermal protection also prevents failure due to creep and fracture, because it keeps the temperature of the ice colder and more stable.

Melting starts when the temperature of the ice surface rises to 0 °C. This does not necessarily mean that the air temperature, t_a, is 0 °C, as evaporation from the ice surface into the air and heat conduction into the bulk ice may considerably cool the surface. In dry air, say at relative humidity of 40%, well-ventilated ice melts only at about +5 °C (Makkonen, 1989). If the bulk ice is much colder than 0 °C the ice surface may not melt even at higher air temperatures.

Conversely, if the ice surface is exposed to direct sunlight, melting may start at temperatures slightly below 0 °C. The significance of this largely depends on the impurities on the surface and on the structure of the ice. A pure, bubble-free ice surface only reflects less than 2% of the incident radiation, but air bubbles in ice increase the reflection up to 50% (Hobbs, 1974). Foreign particles, such as dust, sand and organic materials, decrease the reflection. Thus, during sunny periods, melting of an uncovered ice surface can be effectively reduced by keeping the surface free of contaminants, or by applying white ice (snow, for example) on the surface.

For assessing the need for and proper methods of melt protection it is necessary to estimate various terms of the heat balance on the melting structure. We shall consider a structure whose surface is at the limit of melting: that is, its surface temperature $t_s = 0$ °C. In addition to solar radiation, q_s, the important terms of the heat balance of the ice surface are convective transfer of sensible heat, q_c, convective transfer of latent heat (sublimation), q_e, and warming due to rainwater, q_r. These can be estimated as follows:

$$q_c = h(t_a - t_s) \tag{2.4}$$

$$q_e = \frac{hkL_e}{c_p p_a}(e_q - e_s) \tag{2.5}$$

$$q_r = Rc_w(t_a - t_s) \tag{2.6}$$

where h is the convective heat transfer coefficient, $k = 0.62$, L_e is the latent heat of evaporation, c_p and c_w are the specific heat of air and water, e_s and e_a are the water vapour pressures at the surface (6.1 mb at 0 °C) and in the air, p_a is the air pressure and R is the rainfall intensity.

Precise estimates of the terms are difficult to make, mainly because the convective heat transfer coefficient h depends on wind speed, stratification of the atmospheric boundary layer, and details of the structure.

Nevertheless, it is interesting to make order-of-magnitude estimates under some typical conditions. Let us take an arbitrary example of a location at sea level, where during a warm period the mean wind speed is around 3 m s^{-1}, mean air temperature is +15 °C, mean relative humidity is 80%, mean solar radiation is about 200 W m^{-2} and mean precipitation is 2 mm day^{-1} (i.e. 2.3×10^{-5} kg m^{-2} s^{-1}). Under such conditions, $e_a = 13.6$ mb, and a value of 20 W m^{-2} K^{-1} is a typical estimate for h. Then equations (2.4)–(2.6) give $q_c \sim 300$ W m^{-2}, $q_e \sim 230$ W m^{-2}, and $q_r = 1.5$ W m^{-2}. If we assume that the ice surface reflects 30% of the incident radiation then $q_s = 140$ W m^{-2}.

This arbitrary example shows that the warming due to rain is typically insignificant when compared with the other terms. Transfers of sensible and latent heats are of similar magnitude. In this example the radiation term q_s is about one fifth of the total heating, but at, say, $t_s \sim +3$ °C, it would become the dominant term. In addition to the terms estimated above, there is heat conduction from the ground to the ice. This is of the order of 5–10 W m^{-2}. This is a small fraction of the total heating, but is of some concern because it may affect the foundations of the structure.

The melt rate of the surface, M, is determined by

$$M = \frac{q_m}{L_m} \tag{2.7}$$

where q_m is the heat flux causing the melting and L_m is the latent heat of melting.

Thus it is easy to estimate the corresponding melt rate by the various terms of the heat balance by replacing q_m in equation (2.7) with the appropriate heat flux. This gives 8 cm day^{-1} for sensible heating, 6 cm day^{-1} for condensation-related heating and 4 cm day^{-1} for radiation heating in the above example. The total melt rate in this example is then about 5 m per month.

An efficient way to reduce the heat exchange and melting of an ice structure is to minimize the airflow around it. This will then minimize h in equations (2.4) and (2.5). When possible, the structure should be sheltered from the direct effect of the wind, thereby reducing the airflow down to the free convection that occurs owing to temperature gradients (and consequent air density gradients) only. Moreover, the free convection can be made very small by surrounding the structure by walls. This will produce an effect similar to the uncovered freezer boxes in food stores; as cold air is denser than warm air, there is hardly any free convection if the downflow of the cold air is prevented by walls. Accordingly the best site for an ice structure is in a pit.

In addition to reducing the convection of air, which is the main cause of melting, it is of course possible to use heat insulation materials for melt protection. With modern insulation materials or even with sawdust (Fitch and Jones, 1973) the k-values for a relatively thin layer ($\sim$ 10 cm) are of the order of 0.2. Such a protection would result in $q_c = 3$ W m^{-2} in the above example. The insulation material also reduces the condensation flux and – if the outer surface is made reflective – it can almost remove radiative heating. Using such a cover the total heat flux can easily be reduced down to $q_m < 5$ W m^{-2} and the corresponding melt rate to about 1 mm day^{-1}: that is, 3 cm per month in our arbitrary example. With thicker coatings (and more cost) melting can be further reduced.

Figures 2.3–2.5 present examples of melt protection tried in the field. It is evident from the figures that practical problems exist in covering ice structures at a large scale. Fixing the covers effectively is essential and rather difficult because of the wind. One must also consider flow routes for rainfall and stresses by ice deformation and melt pools (Fig. 2.5).

Fig. 2.3 Snow storage for summer use. City of Tokamachi, Japan.

Fig. 2.4 Melt protection covers on an artificial ice island. (Photo: Exxon Production Research Co.)

Summer deterioration of offshore grounded ice masses is typically dominated by wave action rather than surface abrasion. Heat insulation covers can also be applied to the edges of underwater parts of structures (Fitch and Jones, 1973; Poplin *et al.*, 1991), as shown in Fig. 2.4. Again, effective fixing of the covers is the major technical problem.

References

Fitch, J.L. and Jones, L.G. (1973) *Method of forming and maintaining offshore ice structures.* US Patent No. 3,750,412, 14 pp.

Fig. 2.5 Melt-water pools on insulation covers on an ice island. (Photo: Exxon Production Research Co.)

Hobbs, P.V. (1974) *Ice Physics*, Clarendon Press, Oxford, 837 pp.

Makkonen, L. (1989) Estimation of wet snow accretion on structures. *Cold Regions Science and Technology*, **17**, 83–8.

Poplin, J.P., Weaver, J.S., Gutali, K.C., Lord, S. and Sisodiya, R.G. (1991) Experimental field study of spray ice ablation, in *Proceedings of 11th International Conference on Port and Ocean Engineering under Arctic Condiditons (POAC)*, St John's, Newfoundland, Vol. 1, pp. 259–72.

2.5 CONSTRUCTION TECHNIQUES

2.5.1 Use of natural ice

Close to ice-covered lakes, rivers and coastal areas, ice for use as a construction material is available in mid-winter. Cutting and transportation of block ice to the construction site is required. These operations are straightforward, and cutting machines for this purpose exist (Mellor, 1986; Bogorodsky *et al.*, 1987). Ice blocks can be used as large bricks. They need to be frozen together, because the very low friction of ice may otherwise cause sliding of the contacts. Freezing the blocks together is easily done either by melting a very thin layer at the interface or by spraying small

Fig. 2.6 A structure made of block ice (Photo C. A. Wortley)

quantities of water. If the blocks are lifted straight from the water they should be placed upside down. This will assist in forming a good contact, because the colder side of the ice will effectively freeze the wet interface.

So far, block ice has been rarely utilized in construction. Its potential has, however been demonstrated in various ice festivals (Figs 2.6–2.10).

Ice blocks and other kinds of structural member can also be made out of snow. Compacting wet snow or careful construction by simultaneous heating techniques result in rigid ice used to build snow sculptures and statues (Fig. 2.11). Buildings measuring 40 × 15 × 25 m have been made this way at the Sapporo snow festival in Japan. Other conventional construction materials can, of course, be used to reinforce the critical parts of structures made of block ice or snow.

Fig. 2.7 Snow building in China. (Photo: J. Lillberg)

Fig. 2.8 Snow building in China. (Photo: J. Lillberg)

It is also possible to utilize natural ice in making roads and runways. Glacier ice, for example, can be cleaned and the natural large-scale roughness removed. In areas with thick snow-cover it is more feasible to compact snow to make roads or runways. Both techniques have been extensively developed, and described in recent review papers (Mellor and Swithinbank, 1989; Abele, 1990).

Fig. 2.9 Snow building in China. (Photo: J. Lillberg)

Fig. 2.10 Snow building in China. (Photo: J. Lillberg)

2.5.2 Flooding

During cold periods it is easy to make an ice sheet thicker by pumping water onto the ice. This is usually done in steps, so that the water layer covers the required area and a new layer is pumped after the previous one has frozen. Compacted snow or other material can be used as walls to keep the water in the construction area. In principle, the same method can be used to make ice in any shape by using moulds filled with water. This

Fig. 2.11 Snow building at Tokamichi Snow Festival, Japan.

requires that the expansion of water upon freezing is taken into account either by drainage valves or by flexible moulds.

The advantages of flooding to make ice are the relatively small amount of manual labour, simple equipment and the good quality of the ice (Smorygin, 1988). The main problem is the slow rate of ice formation. Examples are given by Nakawo (1980), for example, showing that in good conditions (air temperature ~ −30 °C, wind speed ~ 4 m s⁻¹) a growth rate of about 10 cm per day is possible. This is usually insufficient for building massive structures. Theoretical methods explained in section 2.4 can be used to estimate the growth rates in various applications.

2.5.3 Spraying

The origin of spraying in making ice is in the development of snow gun technology (these are used to make snow in skiing resorts). The effectiveness of the method is based on the fact that heat transfer between water and air is approximately proportional to the surface area of the interface. By spraying small drops in air this surface area is made much larger, thereby boosting the cooling and freezing of the water.

As in natural spray icing (Chapter 4), the changes in the properties of a cloud of spray droplets during its flight are an important factor. Analyses of these changes have been made (Sosnovskii, 1980; Zarling, 1980; Sackinger and Sackinger, 1985; Sosnovskii, 1987; Szilder and Lozowski, 1988), but owing to the very complicated nature of droplet flight, dispersion and cooling, there is still a lot of work left to do in this area. Such factors as water flux and pressure, nozzle type, spraying angle, wind speed and air temperature all play a role in the process. Spraying techniques normally cause the drops to partly freeze while they are in air (Forest *et al.*, 1992). Finally, it should be mentioned that the efficiency of artificial spray ice-making may be improved by additives. Ice-nucleating bacteria are already used to improve artificial snow-making in some skiing resorts. Additives that reduce the surface tension of water and thereby result in smaller spray droplets have also been tested (Pare *et al.*, 1986).

Spraying technology has developed rapidly, and at present large high-pressure fire-fighting nozzles are mostly used in producing the spray. The effectiveness of spray ice construction and the initial properties of spray ice can be theoretically estimated along the lines of Chapter 4 of this report. The main difference is that artificial spray ice usually grows on a horizontal rather than on a vertical surface. This requires a different approach for estimation of the heat transfer coefficient h. Further analysis can be found in Szilder and Lozowski (1988), Allyn and Masterson (1989), Szilder and Lozowski (1989) and Shatalina (1990).

Spray ice has been suggested for constructing various types of structure, and feasibility studies have been made (Glockner, 1986; Titneva, 1986; Smorygin, 1988; Kärnä *et al.*, 1992). In practice, the use of man-made sea-spray ice has been limited to constructing grounded or floating artificial ice masses offshore. In some cases these masses have been aimed at protecting an existing structure from the impact of floating sea ice (Finncase and Johns, 1987; Kemp *et al.*, 1988). Artificial ice islands for exploratory drilling have been made by spraying (Cox and Utt, 1986; Goff and Masterson, 1986; Prodanovic, 1986; Savel'ev *et al.*, 1986; Faiko, 1990), and this has recently turned out to be the most useful application of spray ice construction.

The mechanical properties of spray ice depend on various factors and differ from those of natural or flooded ice discussed in section 2.2. The material behaviour can be generally described as nonlinear viscoelastic–plastic, and its shear strengths are temperature- and rate-dependent

(Weaver and McKeown, 1986). Typical shear strengths measured in a spray ice island are of the order of 0.7 MPa (Cox and Utt, 1986; Allyn and Masterson, 1989), but there is a strong dependence on, for example, the salinity of the ice (Laforte and Lavigne, 1991) and the settlement time (Vinogradov and Masterson, 1989; Steel *et al.*, 1990). Compressibility of spray ice increases rapidly with increasing density and particle size (Domaschuk *et al.*, 1992). Thus the construction methods need to be carefully controlled in order to optimize ice properties.

2.5.4 Artificial freezing

In some applications – for instance, when natural cold air is not available – one might consider artificial freezing of water to make construction material. This method is technically possible and is, of course, generally used at a small scale to make ice cubes for drinks and ice for ice skating. The feasibility of this method to build large structures is limited by the energy requirement. The latent heat of freezing water is 3×10^8 J m^{-3}, which means that making, say, a $20 \times 20 \times 20$ m cubic ice mass requires 2.4×10^{12} J of energy. This is equivalent to 6.6×10^5 kWh, resulting in a energy cost of US$33,000 at a typical cost of 5 cents/kWh, and assuming no energy losses in the process. This calculation gives an approximate minimum price of US$4 per cubic metre: still a reasonable figure when compared with other building materials.

A normal refrigeration apparatus, however, is a more expensive investment, and this is perhaps the main reason for not developing these techniques further. One may, for example, estimate the power requirement for making the above-mentioned ice mass in a month. This gives 6.6×10^5 kWh/$(30 \times 24$ h$) \approx 1$ MW. The cost of such a cooling machinery is typically around US$1 million. Natural heat pipes may reduce this investment significantly.

References

Abele, G. (1990) *Snow Roads and Runways*. US Army Cold Regions Research and Engineering Laboratory Monograph 90-3, 100 pp.

Allyn, N. and Masterson, D. (1989) Spray ice construction and simulation, in *Proceedings of 8th International Conference on Offshore*

Mechanics and Arctic Engineering (OMAE), The Hague, Netherlands, ASME, Vol. IV, pp. 253–62.

Bogorodsky, V.V. Gavrilo and Nedoshivin. (1987) *Ice Destruction – Methods and Technology*, D. Reidel Publ. Co., Holland, 214 pp.

Cox, G.F.N. and Utt, M.E. (1986) Ice properties in a grounded man-made ice island, in *Proceedings of 5th International Conference on Offshore Mechanics and Arctic Engineering (OMAE)*, Tokyo, ASME, Vol. IV, pp. 135–42.

Domaschuk, L., Shields, D.H. and Tong, Y.X. (1992) Compressibility of spray ice, in *Second IOPE Conference*, Vol. II, pp. 651–5.

Faiko, L.I. (1990) Experience of construction large ice structures. *Acad. Nauk USSR, Data of Glaciological Studies*, **68**, 104–6 (in Russian).

Finncase, R.G. and Johns, H.O. (1987) *Ice barrier construction*. US Patent No. 81 4,523,879, 18 pp.

Forest, T.W., Szilder, K. and Lozowski, E.P. (1992) Ice island construction with ice-water mixtures. *The Northern Engineer*, **23** (2–3), 19–27.

Glockner, P.G. (1986) Reinforced ice domes as temporary enclosures for cold regions, in *Proceedings of 7th International Conference on Offshore Mechanics and Arctic Engineering (OMAE)*, ASME, Vol. IV, pp. 105–110.

Goff, R.D. and Masterson, D.M. (1986) Construction of a sprayed ice island for exploration, in *Proceedings of 5th International Conference on Offshore Mechanics and Arctic Engineering (OMAE)*, Tokyo, ASME, Vol. IV, pp. 105–10.

Kärnä, T., Hakala, R. and Makkonen, L. (1992) *A layered supporting structure made of ice*. Finnish Patent No. FI-85173, 15 pp. (in Finnish)

Kemp, T.S., Foster, R.J. and Stephens, G.S. (1988) Construction and performance of the Kadluk 0-07 sprayed ice pad, in *Proceedings of 9th POAC Conference*, Vol. III, pp. 551–64.

Laforte, J-L. and Lavigne, L. (1991) Microstructure and mechanical properties of ice accretions grown from supercooled water droplets containing NaCl in solution, in *Proceedings of 3rd International Workshop on Atmospheric Icing of Structures*, Vancouver, pp. 149–53.

Mellor, M. (1986) *Equipment for making access holes through arctic sea ice.* US Army Cold Regions Research and Engineering Laboratory Special Report 86-32, 34 pp.

Mellor, M. and Swithinbank, C. (1989) *Airfields on Antarctic glacier ice.* US Army Cold Regions Research and Engineering Laboratory Report 89–21, 97 pp.

Nakawo, M. Heat, (1980) Exchange at surface of built-up ice platform during construction. *Cold Regions Science and Technology*, **3**, 323–33.

Pare, A., Carlson, L.E., Bourns, M. and Karim, N. (1986) The use of additives in sprayed sea water to accelerate ice structure construction, in *Proceedings of International Symposium on Cold Regions Heat Transfer*, Edmonton, pp. 123–29.

Prodanovic, A. (1986) Man-made ice island performance, in *Proceedings of 5th International Conference on Offshore Mechanics and Arctic Engineering (OMAE)*, Tokyo, ASME, Vol. II, pp. 89–95.

Sackinger, W.M. and Sackinger, P.A. (1985) *The freezing of sprayed sea water to produce artificial spray ice.* Unpublished report, Taylor Naval Ship R.D. Center, Annapolis, USA, 91 pp.

Savel'ev, B.A., Gagarin, V.E. and Pizhaskov, N.H. (1986) Structure and mechanical properties of artificial ice, in *Formation of Frozen Rocks and Forecasts of Cryogenic Processes* (ed. T.M. Kaplina), Moskow, Nauka, pp. 169–76 (in Russian).

Shatalina, I.N. (1990) *Technology of ice massif buildings.* Interdepartmental collected scientific papers (edited by N.F. Krivonogova), Leningrad VNIIG, pp. 92–100 (in Russian).

Smorygin, G.I. (1988) *Theory and Methods of Making Artificial Ice*, Nauka, Novosibirsk, 282 pp. (in Russian).

Sosnovskii, A.V. (1980) Mathematical modelling of the ice formation process in a plume of artificial rain. *Acad. Nauk USSR, Data of Glaciological Studies*, **38**, 49–54 (in Russian).

Sosnovskii, A.V. (1987) Analysis of the computation methods of ice formation in a spray cone, *Acad. Nauk USSR, Data of Glaciological Studies*, **59**, 61–8 (in Russian).

Steel, A., Morin, P.J. and Clark, J.I. (1990) Behaviour of laboratory-made spray ice in triaxial compression testing, *Journal of Cold Regions Engineering*, **4** (4), 192–204.

Szilder, K. and Lozowski, E.P. (1988) The influence of meteorological variables on the construction of ice platforms, in *Proceedings of 9th International Association for Hydraulic Research (IAHR) Ice Symposium*, Sapporo, Japan, Vol. II, pp. 410–24.

Szilder, K. and Lozowski, E.P. (1989) A time-dependent thermodynamical model of the build-up of ice platforms. *Journal of Glaciology*, **35**, 169–78.

Titneva, G.A. (1986) Construction of temporary ice crossings and winter roads. *Energeticheskoe Stroitelistvo*, **6**, 13–14 (in Russian).

Weaver, J.S. and McKeown, S. (1986) Observations on the strength properties of spray ice, in *Proceedings of 5th International Conference on Offshore Mechanics and Arctic Engineering (OMAE)*, Tokyo, ASME, Vol. II, pp. 96–104.

Vinogradov, A.M. and Masterson, D.M. (1989) Time dependent settlement of ice and earth offshore structures, in *Proceedings of 8th International Conference on Offshore Mechanics and Arctic Engineering (OMAE)*, The Hague, Netherlands, ASME, Vol. IV, pp. 263–68.

Zarling, J.P. (1980) *Heat and mass transfer from freely falling drops at low temperatures*. US Army Cold Regions Research and Engineering Laboratory Report 80–18.

3

Construction on ice

3.1 BEARING CAPACITY OF ICE

3.1.1 General considerations

For years the ice covers of rivers and lakes have been used for transportation and water-crossing purposes. More recently, offshore drilling operations in the Arctic regions have considerably increased the interest in the utilization of ice covers for winter roads, airfields and the support of material and equipment for construction. Thickened ice platforms have been used for setting offshore drilling oil rigs. The loading condition can last from a few seconds to weeks and even months.

Predicting the bearing capacity of the ice cover requires very good information on the temperature- and time-related viscoelastic and strength properties of the ice in a complex numerical analysis. However, the theory of linear elasticity based on nominal values for Young's modulus and flexural strength seems to work reasonably well in most practical cases. Observations on deflections and cracking of a loaded ice cover will naturally make this kind of analysis more reliable. Because there are many unknown factors, conservative assumptions should be used, especially when working safety is to be considered.

3.1.2 Dependence on ice thickness

The bearing capacity of floating ice covers depends on many things. One of the most important things to be considered is the effective ice thickness. This is usually taken as the thickness of a uniform good-quality columnar-grained ice layer. Columnar ice is formed by the freezing of water. It is also called **clear blue ice** and is generally the strongest ice type. The other type of ice is **granular ice** or **snow ice**, which is formed by the freezing of water-saturated snow. The strength of snow ice, which has a relatively high air content, depends on the density. High-density snow ice can be almost as strong as clear blue ice, but it is normally considered to be half as strong as clear blue ice (*Handbook of Occupational Health and Safety*, 1982). When the colour of ice is grey, this generally indicates melting between the grain boundaries. This kind of ice must be considered with extra caution as a load-bearing surface.

When the temperatures are not low enough to obtain the necessary natural ice thickness, or when snow cover decreases the ice growth rate, the ice thickness can be increased by flooding (adding water on top of the existing ice cover). The period of a flooding-and-freezing cycle for a given layer thickness has to be long enough to achieve the required degree of solidification of the new ice layer (Nakawo, 1983). To achieve maximum strength, any snow cover should be removed before each flooding operation. During warm temperatures the boundary between the layers may very easily loose the capability to transfer shear and the layers may begin to act separately.

If the ice cover is assumed to behave elastically under load and if the shear deformations are neglected, the governing differential equation for the ice cover is given by

$$v^4 w = \frac{q - kw}{D} \tag{3.1}$$

where D is the plate rigidity $= Eh^3/12(1 - v^2)$, q is the applied distributed load, k is the subgrade reaction $= 9.81 \ kN/m^2$, E is Young's modulus, v is Poisson's ratio, h is the thickness of the ice cover, and w is the deflection.

Radial and tangential moments and stresses for a uniform load q distributed over a circular area of radius a can be derived from Wyman's (1950) equations. Maximum stress under the load is given by

$$\sigma_{max} = \frac{3qaD \ kei'\left(\dfrac{a}{L}\right)(1 + v)}{kh^2 L^3} \tag{3.2}$$

where $L = [Eh^3/12k(1 - v^2)]^{1/4}$, and kei' is the first derivative of the modified Bessel function kei.

When the ratio a/L is small, the deflection under load can be approximated by

$$W_{max} = \frac{q\pi a^2}{8kL^2} \tag{3.3}$$

In practice, the loading condition is often more complicated than described. The ice cover can already have cracks, the loading may consist of several distributed loads with various forms, and the vicinity of structures or shores may have some effect on the maximum stress in the ice cover.

Because the theory of thin plates is used, the maximum stress in an uncracked ice sheet is overestimated when the radius a of the loaded area is small. On the other hand, the possibility of punch-through failure becomes evident for highly concentrated loads, especially if the ice temperatures are high and the loading static.

For design purposes it has been usual to assume a very simple formula of the type

$$P = Ah^2 \tag{3.4}$$

where P is in kg, h is in m, and A is a constant that is determined from experience and the conditions of the ice cover (Michel, 1978; Gold, 1981). The usual recommended range for loads to be placed on ice of thickness h is $P = 3.5 \times 10^4$ to $7.0 \times 10^4 \, h^2$. The lower limit should be used for uncontrolled situations. Greater values for A can be used for good-quality ice covers in the case of a moving load or if risk is acceptable.

3.1.3 Dependence on ice properties

Ice in ice bridges and platforms is subjected to a wide variation in structure and quality, although careful control may have been exercised in construction. The strength of the ice cover depends on ice structure, but also, in certain cases, on temperature. Short-term flexural and shear strengths of freshwater ice cover are relatively independent of temperature below freezing conditions (Gow and Ueda, 1989). However, considerable drops in air temperature can temporarily cause internal stress in an ice cover and reduce its bearing capacity. This can often occur during overnight periods, when the temperature is much lower than the preceding average for the day.

The removal of snow from an ice cover during periods of low temperature has the same kind of effect as a marked temperature drop. At temperatures close to and above the freezing point, ice begins to lose its strength rapidly. The ice will thaw in the sunlight, but a clean snow layer on the ice cover will reduce significantly the solar radiation penetrating the cover. The flexural and shearing strengths of sea ice depend on brine volume, which is a function of temperature. A reduction of bearing capacity of 10% for each degree-day of temperatures above −1 °C can be used as a rule of thumb. For saline ice more serious restrictions are required, as they are for long-term loads.

Because of the viscoelastic behaviour of ice, the bearing capacity of floating ice covers depends on the loading time. Also, the type and the area of loading have to be taken into consideration. Depending on the loading conditions three types of bearing capacity problem can be distinguished: dynamic, static and quasi-static (Kerr, 1976).

Dynamic problems deal primarily with moving loads on ice covers used for transportation, such as aircraft runways, ice roads and crossings. In this case the analysis of the problem depends upon the velocity of the moving loads (Beltaos, 1978; Kerr, 1983). Repeated oscillatory forces may cause a decrease in the bearing capacity due to fatigue (Haynes *et al.*, 1993).

Static problems usually involve comparatively large loads and are mainly concerned with the bearing capacity of ice plates in terms of ultimate failure. When the loads are for only a few seconds on the ice cover, the ice will behave elastically. If ice failure happens during short-term loading, failure conditions are attained instantaneously at the time of the load application.

Loads of long duration (**quasi-static** bearing capacity problems) may not cause an instantaneous breakthrough. The ice cover deforms at first elastically, and then continues to deform in creep. Creeping will happen especially in the vicinity of load (Fig. 3.1a).

Depending on the initial stress in the ice cover, the resulting time–displacement graphs may be of the type shown in Fig. 3.2. In the case of the upper curve the ice cover is able to carry the load at the beginning but after some time the ice cover will creep very quickly and the load breaks through the cover. According to tests made by Assur (1961) the magnitude of the load that can be safely parked on an ice cover decreases, the longer the time it is required to stay on the ice (Fig. 3.2b).

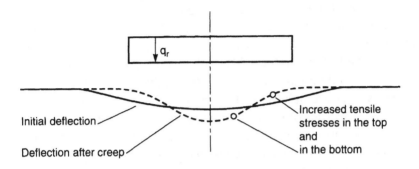

Fig. 3.1 Change of deflection profile due to creep (Fransson and Elfgren, 1986).

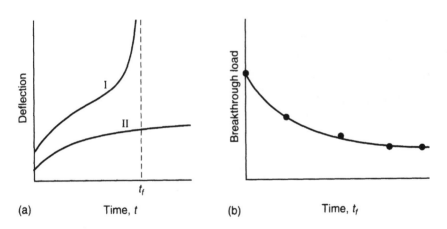

Fig. 3.2 Two typical time–displacement curves for ice cover under quasi-static loading conditions.

   The ice cover usually has many cracks, which may not necessarily indicate a reduction in the load-bearing capacity of the cover. If the crack-opening rate is noticeable, or the crack is wet, loads should be reduced. Special attention should be given if two wet cracks meet at right angles. Also, the position of cracks should be noticed.

3.2 TEST METHODS FOR BEARING CAPACITY

All common methods of determining the bending strength and bulk modulus of elasticity for ice are based on the elastic theory of homogeneous, isotropic materials. Ice is neither isotropic nor homogeneous, and behaves as an elastic material only when the rate of deformation is sufficiently high. It is well known that different testing methods yield different estimates of bending strength σ_b and, especially, bulk modulus E_b. Such estimated values are quite useful, as they can be applied to compare the mechanical properties of ice at different locations in the field.

Possibly the most common testing method, both in the field and in the laboratory, is the *in-situ* **cantilever-beam test**, because of its relative simplicity. The measurement of beam deflection during beam loading may often be a difficult task to perform with sufficient reliability, especially in the field. In Table 3.1 (Schwartz *et al.*, 1981) equations are given for the determination of the flexural strength and the strain modulus for the case of cantilever, three-point-loaded and four-point-loaded beams. The variables in the table are defined as:

P' = maximum force required to break the beam
P = load applied to beam
δ = corresponding beam deflection
b, h, l = beam width, thickness and length

Table 3.1 Equations for the flexural strength of brackish water ice by *in-situ* tests (from Schwartz *et al.*, 1981)

	Cantilever beam	Three-point loading	Four-point loading
Bending strength	$6\dfrac{P'l}{b\,h^2}$	$\dfrac{3}{2}\dfrac{P'l}{b\,h^2}$	$3\dfrac{P'c}{b\,h^2}$
Effective modulus	$\dfrac{4}{b}\left(\dfrac{l}{h}\right)^3\dfrac{P}{\delta}$	$\dfrac{1}{4b}\left(\dfrac{l}{h}\right)^3\dfrac{P}{\delta}$	$\dfrac{3}{2b}\left(\dfrac{l-2c}{h}\right)^3\dfrac{c}{l-2c}\dfrac{P}{\delta}$

An *in-situ* beam test has the advantage of being relatively straightforward to perform for at least three reasons:

1. Load is applied to a relatively large sample of ice.
2. In the case of sea ice the problem of brine drainage is avoided.
3. The natural temperature gradient is maintained.

It is important to do complete reporting of experimental procedures and test conditions. At least the following points should be considered in carrying out beam test programmes:

1. beam geometry;
2. stress concentration (Frederking and Svec, 1985);
3. loading rate (Määttänen, 1975);
4. instrumentation;
5. test conditions;
6. beam preparation.

The deflections and stresses of floating ice covers can also be measured by loading the ice plate. The ice plate can be loaded to breakthrough condition, or the behaviour of the ice can be studied during long-term loading. Both tests have been reported in the literature (Kingery, 1962; Fransson, 1984).

3.3 REINFORCEMENT OF ICE COVERS

3.3.1 General

The bearing capacity of an ice cover can be improved by the following three methods:

1. increasing the ice thickness;
2. changing the ice properties;
3. introducing reinforcement.

All three methods have their merits and drawbacks. Increasing the thickness is often the easiest way. However, in mild weather flooding in order

All three methods have their merits and drawbacks. Increasing the thickness is often the easiest way. However, in mild weather flooding in order to build up thicker ice can be very time-consuming. On rivers a thick ice road may also cause erosion downstream. Changing the ice properties by, for example, sand, sawdust or chemicals may also be insufficient in such a situation, and the best way to construct a strong platform may be to introduce reinforcement. The cost of the additional material (the reinforcement) can be balanced by a shorter construction time and by a better guarantee that the required load-carrying capacity will be obtained.

In Northern Sweden, for example, reinforced ice covers have been used as platforms for piledrivers and bucket excavators during bridge construction works. This has facilitated the works and has provided the builders with an economic way of carrying out piling and excavation. Also, during repair work and when an old structure is being demolished, ice covers have been reinforced to serve as strong platforms (Fransson, 1984).

3.3.2 Reinforcement methods

The positive effects of reinforcement on ice beams have been known for a long time (Gold, 1989). During the Second World War the feasibility of building ships from ice was studied. When wood or steel was placed in the bottom of ice beams the loading capacity became 1.5–2 times as large as for the unreinforced beams. The loading capacity for manufactured ice was further increased by adding different fibres to the water (Table 3.2).

Table 3.2 Relative increase in modulus of rupture: beams $5 \times 10 \times 89$ cm, reinforced by various materials; temperature: $-12°C$ to $-18°C$ (Gold, 1989)

Average modulus of rupture of plain ice: S_i = 1.7 MPa
Peat moss $0.81 \, S_i$
Straw $2.70 \, S_i$
Sawdust $1.86 \, S_i$
Flax straw $3.44 \, S_i$
Wood shavings $1.93 \, S_i$
Blotting paper $4.06 \, S_i$
Hay $2.15 \, S_i$
Absorbent cotton $9.33 \, S_i$
Pykrete (6.5% wood pulp) $2.8 \, S_i$
Pykrete (10.1% wood pulp) $4.35 \, S_i$

Ice reinforced with a polymeric mesh (geogrid) has been reported to have 10–38% higher bearing capacity than natural ice (Haynes *et al.*, 1992). The effect is more significant the thinner the ice. The flexural strength of reinforced cantilever beams has been studied by Den Hartog and Ohstrom (1976) and Fransson (1979). With reinforcement bars of wood or steel frozen to the surface of lake ice the bending moment was usually raised 2–3 times compared with plain ice. Fig. 3.3 shows the load –displacement curves of reinforced ice beams. The cantilever beams were pushed downwards, resulting in tension of the embedded reinforcement bars.

It was found that less than 1% of wooden bars was needed to achieve stronger ice. If the content of reinforcement was greater than 1% the ice was likely to fail by shearing. Also, the weak bonding between layers of flooded ice sometimes limited the load capacity.

Thick ice plates reinforced with wooden bars placed in a square pattern were tested in field experiments (Fransson, 1983, 1984). A 0.6 m thick re-inforced ice plate could not be cracked because of insufficient stroke capacity of the loading equipment. At a central deflection of 100 mm the load was 560 kN and no cracks were observed on the ice surface. The remaining deformation immediately after unloading was 20 mm. The

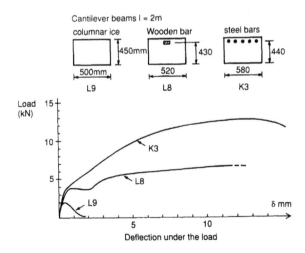

Fig. 3.3 Load-deflection for ice cantilever beams with various amounts of reinforcement

effects of reinforcement placed at the upper surface of an ice sheet were summarized as follows.

1. It protects ice against thermal shock.
2. The remaining deformation becomes smaller.
3. Reinforced ice sheets have a higher load-bearing capacity, particularly in the vicinity of an ice edge or a wet crack.

The creep behaviour of reinforced ice beams has been studied by Cederwall (1981). It was shown that the flexural creep rate was reduced and started at lower tensile stresses when the reinforcement was activated by cracking.

The creep deformation of reinforced ice plates has also been measured in full-scale tests (Fransson and Elfgren, 1986). A thin layer of saturated frozen sand on top of lake ice was found to be equally effective as wood in this respect but the dark surface was more sensitive to solar radiation.

REFERENCES

Assur, A. (1961) Traffic over frozen or crusted surfaces, in *Proceedings of First International Conference on Mechanics of Soil– Vehicle Systems*, Edizioni Minerva Technica, Torino, Italy, pp. 913–23.

Beltaos, S. (1978) Field studies on the response of floating ice sheets to moving loads. *Workshop on the Bearing Capacity of Ice Covers*, 16–17 October 1978, Winnipeg, Manitoba. Technical Memorandum No. 123, pp. 1–13.

Cederwall, K. (1981) Behaviour of a reinforced ice cover with regard to creep, in *Proceedings of International Conference on Port and Ocean Engineering under Arctic Conditions (POAC)*, Quebec, Canada, Vol. 1, pp. 562–70.

Den Hartog, S.L. and Ohstrom, M. (1976) *Cantilever beam tests on reinforced ice*. US Army Cold Regions Research and Engineering Laboratory, USA, Report 76–7.

Fransson, L. (1979) *Improved Bearing Capacity of an Ice Sheet with Reinforcement*. Luleå University of Technology, Sweden, Diploma work 1979:002E (in Swedish).

Fransson, L. (1983) Full-scale tests of the bearing capacity of a floating ice cover, in *Proceedings of International Conference on Port and*

Ocean Engineering under Arctic Conditions (POAC), Helsinki, Finland, Vol. 2, pp. 687–97.

Fransson, L. (1984) *Bearing Capacity of a Floating Ice Cover*. Luleå University of Technology, Division of Structural Engineering, Licentiate thesis 1984:126, 137 pp. (in Swedish.)

Fransson, L. and Elfgren, L. (1986) Field investigation of load–curvature characteristics of reinforced ice, in *POLARTECH 86 Conference*, VTT, Helsinki 27–30 October 1986, Vol. 1, pp. 175–96.

Frederking, R. and Svec, O. (1985) Stress-relieving techniques for cantilever beam tests in an ice cover. *Cold Regions Science and Technology*, **11**, 247–53.

Gold, L.W. (1981) Designing ice bridges and ice platforms, in *Proceedings of International Association for Hydraulic Research (IAHR) International Symposium on Ice*, Quebec, Vol. II, pp. 658–701.

Gold, L.W. (1989) The Habbakuk Project – Building ship from ice, in *Proceedings of International Conference on Port and Ocean Engineering under Arctic Conditions (POAC)*, Luleå, Sweden, Vol. 1, pp. 1211–28.

Gow, A.J. and Ueda, H.T. (1989) Structure and temperature dependence of the flexural properties of laboratory freshwater ice sheets. *Cold Regions Science and Technology*, **16**, 249–69.

Handbook of Occupational Health and Safety (1982) 3rd edn, *Safety Guide for Operations over Ice*. TB Guide 5-3, Minister of Supply and Services, Canada.

Haynes, F.D., Collins, C.M. and Olson, W.W. (1992) *Bearing Capacity Tests on Ice Reinforced with Geogrid*. US Army Cold Regions Research and Engineering Laboratory Special Report 92–28, 20 pp.

Haynes, F.D., Kerr, A.D. and Martinson, C.R. (1993) Effect of fatigue on the bearing capacity of floating ice sheets. *Cold Regions Science and Technology*, **21**, 257–63.

Kerr, A. (1983) The critical velocities of a load moving on a floating ice plate that is subjected to in-plane forces. *Cold Regions Science and Technology*, **6**, 267–74.

Kerr, A.D. (1976) The bearing capacity of floating ice plates subjected to static and quasi-stating loads. *Journal of Glaciology*, **17**, 229–68.

Kingery, W.D. (1962) *Summary Report – Project Ice-Way*. Air Force Surveys in Geophysics, 145, Air Force Cambridge Research Laboratories, Office of Aerospace Research, US Air Force.

Määttänen, M. (1975) On the flexural strength of brackish water ice by in situ tests, in *Proceedings of 3rd International Conference on Port and Ocean Engineering under Arctic Conditions*, Fairbanks, Alaska, Vol. 1 pp. 349–59.

Michel, B. (1978) *Ice Mechanics*, Les Presses de l'Université, Laval, Québec, 499 pp.

Nakawo, M. (1983) Criteria for constructing ice platforms in relation to meteorological variables. *Cold Regions Science and Technology*, **6**, 231–40.

Schwarz, J. *et al.* (1981) Standardized testing methods for measuring mechanical properties of ice. *Cold Regions Science and Technology*, **4**, 245–253.

Wyman, M. (1950) Deflections of an infinite plate. *Canadian Journal of Research*, **A28**, 293–302.

4

Accretion of ice on structures

4.1 THE PHYSICS OF ICING

4.1.1 Sources of accreted ice

The formation of ice on land structures may derive from cloud droplets, raindrops, snow, or water vapour. Here the term **cloud droplets** implies droplets in clouds observed locally as fog; they are smaller than raindrops and have a lower fall velocity. The effects of water vapour condensation (hoar frost) have been shown to be usually negligible compared with growth rates of ice from impingement of liquid water droplets and snow particles (Makkonen, 1984a).

At sea (offshore structures and shipping), spray from waves breaking against the structure can also cause icing. Examples of ice problems on structures are shown in Figs 4.1 and 4.2.

4.1.2 Fundamentals of the icing process

A condensation of small ice particles forms owing to particles in the air colliding with the object. These particles can be liquid (usually super-cooled) or solid, or a mixture of water and ice. The maximum rate of icing per unit projection area of the object is then determined by the flux density of these particles. The flux density F is a product of the mass concentration of particles, w, and their velocity v with respect to the object. The rate of icing is obtained as follows:

$$\frac{\mathrm{d}M}{\mathrm{d}t} = \alpha_1 \times \alpha_2 \times \alpha_3 \times w \times v \times A \tag{4.1}$$

Fig. 4.1 Atmospheric ice on TV masts. (Photo: Finnish Broadcasting Co.)

Fig. 4.2 Sea spray ice on superstructures of a vessel. (Photo: Finnlines Ltd)

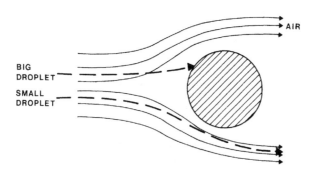

Fig. 4.3 Airstream and droplet trajectories around a cylindrical object.

where A is the cross-sectional area of the object with respect to the direction of particle velocity vector v. The correction factors α_1, α_2 and α_3 vary between 0 and 1 and represent processes that may reduce dM/dt from its maximum value $w \times v \times A$.

In the above equation, α_1 is the **collision efficiency**: that is, the ratio of the flux density of particles hitting the object to the maximum flux density. Its value is below 1, as small particles tend to follow the airstream and may be deflected from their path towards the object as shown in Fig. 4.3.

Factor α_2 is the efficiency of collection of particles hitting the object: that is, the ratio of the flux density of particles that stick to the object to the flux density of particles that hit it. The **sticking efficiency** α_2 is less than 1 when particles bounce off the surface.

Factor α_3 is the **efficiency of accretion**: that is, the ratio of the rate of icing to the flux density of the particles sticking to the surface. When there is no liquid layer or run-off ($\alpha_3 = 1$), the process is called **dry growth** (Fig. 4.4). The ice resulting from dry growth is known as **rime**. The accretion efficiency α_3 is reduced to below 1 when the heat flux from the accretion is too small to cause sufficient freezing to incorporate all the

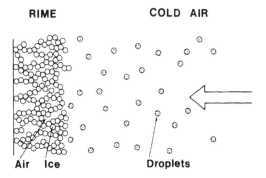

Fig. 4.4 Growth of rime ice (dry growth).

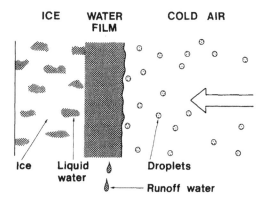

Fig. 4.5 Growth of glaze ice (wet growth).

sticking particles into the accretion. In such a case part of the mass flux of particles is lost from the surface by run-off. A liquid layer on the surface of the accretion and freezing develops beneath this layer (Fig. 4.5). This is **wet growth**. The ice resulting from this process is customarily called **glaze**.

It should be noted that although one speaks of 'icing' and 'icing rate' dM/dt, the accretion that forms may be a mixture of ice and liquid water. In fact, when a liquid film forms at the accretion surface (Fig. 4.5), the growing ice always initially entraps a considerable amount of liquid water (Makkonen, 1987). Liquid water is seldom detected, because the deposits usually freeze completely soon after the icing storm is over.

4.1.3 Rate of icing

Equation (4.1) reveals some of the basic problems of estimating ice loads on structures. Three factors, α_1, α_2 and α_3, which may all vary between 0 and 1, must be determined. In addition, the mass concentration of particles in air, w, the particle velocity v and the cross-sectional area of the object, A, must all be known. Determination of the atmospheric parameters is a meteorological problem, and is not discussed here. The mass concentration w is not discussed here either. Note that w is not a routinely measured parameter; its estimation is a complex problem, and the velocity v is a vector sum of surface velocity, wind speed and the terminal velocity of cloud particles.

The theoretical determination of factors α_1, α_2 and α_3 and A is discussed below.

(a) Collision efficiency

When a droplet moves within the airstream towards the icing object, its trajectory is determined by the forces of aerodynamic drag and inertia. If inertial forces are small, drag will dominate and the droplets will closely follow the stream of air (Fig. 4.3). The actual impingement rate will then be smaller than the flux density of the spray. For large droplets, however, inertia will dominate and the droplets will tend to hit the object without being deflected (Fig. 4.3).

The relative magnitude of the inertia and drag on the droplets depends on the droplet size, the velocity of the airstream, and the dimensions of the icing object. When these are known, the collision efficiency α_2 can be determined theoretically by solving numerically the equations for droplet motion in the airflow. This approach involves numerical solution of the airflow and of the droplet trajectories. The trajectories must be determined for a number of particle sizes and impact positions in order to derive finally the overall collision efficiency α_1. These calculations are complex and computationally costly, but there are several ways in which they can be simplified for practical purposes.

First, the collision efficiency can be parametrized by two dimensionless parameters as follows:

$$K = \frac{\rho_w d^2}{9\mu D} \qquad (4.2)$$

and

$$\phi = \frac{Re^2}{K} \tag{4.3}$$

with the droplet Reynolds number Re based on the velocity v:

$$Re = \frac{\rho_a vd}{\mu} \tag{4.4}$$

where d is the droplet diameter, D is the surface cross-sectional diameter, ρ_w is the water density, μ is the absolute viscosity of air, and ρ_a is the air density.

An empirical fit to the numerically calculated data for α_1 has been developed based on these parameters (Finstad *et al.*, 1988a) as follows:

$$\alpha_1 = A - 0.028 - C(B - 0.0454) \tag{4.5}$$

where

$A = 1.066K^{-0.00616} \exp(-1.103K^{-0.688})$
$B = 3.641K^{-0.498} \exp(-1.497K^{-0.694})$
$C = 0.00637(\phi - 100)^{0.381}$

Second, it has been shown that with good accuracy a single cloud drop parameter – the **median volume diameter** (MVD) – can be used in the calculations without having to calculate α_1 separately for each droplet size category (Finstad *et al.*, 1988b).

The collision efficiency α_1 depends strongly on the particle size, and for a sufficiently large MVD one can assume $\alpha_1 = 1$ in practical applications, unless the structure is extremely large. Therefore α_1 usually needs calculating only when icing is caused by cloud droplets. In cases of precipitation (both rain and snow as well as sea-spray icing), the collision efficiency is close to 1.

(b) Sticking efficiency
When a supercooled water drop hits an ice surface, it freezes rapidly and does not bounce (Fig. 4.4). If there is a liquid layer on the surface the droplet spreads on the surface and again there is no bouncing (Fig. 4.5). Small droplets that leave the surface may be created in such instances ow-

ing to splintering, but their relative volume is too small to have any significant effect on icing. Therefore liquid water droplets can generally be considered not to bounce: that is, for water droplets $\alpha_2 = 1$.

Snow particles, however, bounce very effectively. For completely solid particles (dry snow), the sticking efficiency α_2 is basically 0, but where the particles have a wet layer on their surface sticking is more effective. At small impact speeds and favourable temperature and humidity conditions, α_2 is close to unity for wet snow.

Currently there is no theory for the sticking efficiency of wet snow. The available approximation methods for α_2 are empirical equations based on laboratory simulations and some field observations. The best first approximation for α_2 is probably as follows (Admirat *et al.*, 1986):

$$\alpha_2 = \frac{1}{v} \tag{4.6}$$

where the wind speed v is in m s^{-1}. When $v < 1$ m s^{-1}, $\alpha_2 = 1$.

Air temperature and humidity also affect α_2, but at present there are insufficient data to take them into account. However, it should be noted that $\alpha_2 < 0$ only when the snow particle surface is wet, so that for snow $\alpha_2 = 0$ when the wet-bulb temperature is below 0 °C (Makkonen, 1988b).

(c) Accretion efficiency
In dry-growth icing, all the impinging water droplets freeze, and the accretion efficiency $\alpha_3 = 1$. In wet-growth icing, the freezing rate is controlled by the rate at which the latent heat released in the freezing process can be transferred away from the freezing surface. The portion of the impinging water that cannot be frozen by the limited heat transfer runs off the surface owing to gravity or wind drag (Fig. 4.5).

For wet-growth icing the heating balance on the icing surface can be defined as

$$Q_f + Q_v = Q_c + Q_e + Q_l + Q_s \tag{4.7}$$

where Q_f is the latent heat released during freezing, Q_v is the frictional heating of air, Q_c is the loss of sensible heat to air, Q_e is the heat loss due to evaporation, Q_l is the heat loss (gain) in warming (cooling) impinging water to freezing temperature, and Q_s is the heat loss due to radiation.

The terms of the heat balance equation can be parametrized using the meteorological and structural variables.

The heat released during freezing is transferred from the ice–water interface through the liquid water into the air. Such supercooling results in a dendritic growth morphology, and consequently some liquid water is trapped within the spray ice matrix. As the unfrozen water can be trapped without releasing any latent heat, the term Q_f in equation (4.7) is defined as

$$Q_f = (1 - g)\alpha_3 FL_f \tag{4.8}$$

where L_f is the latent heat of fusion, g is the liquid fraction of the accretion, and F is the flux density of water to the surface ($F = \alpha_1\alpha_2 Wv$).

Attempts to determine the liquid fraction have been made both theoretically (Makkonen, 1990) and experimentally (Gates *et al.*, 1986). These studies suggest that it is rather insensitive to growth conditions, and the value $g = 0.26$ is a reasonable first approximation (Makkonen, 1987).

The kinetic heating of air, Q_v, is a relatively small term, but as it is easily parametrized as follows:

$$Q_v = \frac{hrv^2}{2c_p} \tag{4.9}$$

it is usually included in the heat balance equation. Here h is the convective heat transfer coefficient, r is the recovery factor for viscous heating, v is the wind speed, and c_p is the specific heat of air.

The convective heat transfer is

$$Q_c = h(t_s - t_a) \tag{4.10}$$

where t_s is the temperature of the icing surface and t_a is the air temperature.

The evaporative heat transfer is parametrized as

$$Q_e = \frac{h\varepsilon L_e(e_s - e_a)}{c_p p} \tag{4.11}$$

where ε is the ratio of the molecular weights of dry air and water vapour ($\varepsilon = 0.622$), L_e is the saturation water vapour pressure over the accretion surface, e_a is the ambient vapour pressure in the airstream, and p is the air pressure.

Here, e_s is a constant (6.17 mb) and e_a is a function of the temperature and relative humidity of ambient air. The relative humidity in a cloud is usually assumed to be 100%.

Heat loss (or gain) Q_1 is caused by the temperature difference between the impinging spray droplets and the surface of the icing object.

$$Q_1 = Fc_w\left(t_s - t_d\right) \tag{4.12}$$

where c_w is the specific heat of water, and t_d is the temperature of the droplets at impact.

For cloud droplets we can assume that $t_d = t_a$. This assumption is also usually made for supercooled raindrops. For atmospheric icing $t_s = 0$ °C, but for sea-spray icing, where the water contains salt, t_s must be modelled separately (Makkonen, 1987).

The heat loss due to long-wave radiation may be parametrized as

$$Q_s = \delta a\left(t_s - t_a\right) \tag{4.13}$$

where δ is the Stefan–Boltzmann constant (5.67×10^{-8} W m^{-2} K^{-4} and a is the radiation linearization constant (8.1×10^7 K^3). This equation only takes into account long-wave radiation and assumes emissivities of unity for both the icing surface and the environment.

Using the parametrizations of equations (4.8)–(4.13) in the heat balance equation (4.7) and solving the accretion fraction gives the following equation:

$$\alpha_3 = \frac{h}{F(1-g)L_f}\left[\left(t_s - t_a\right) + \frac{\varepsilon L_e}{c_p p}\left(e_s - e_a\right)\frac{rv^2}{2c_p}\right] + \frac{c_w\left(t_s - t_d\right)}{\left(1 - L_f\right)g} + \frac{\delta a\left(t_s - t_a\right)}{F(1-g)} \tag{4.14}$$

As to the heat transfer coefficient h in equation (4.14), there are standard methods for estimating both local and overall values of this for smooth objects of varying size and shape, but in most icing models the values for cylinders are considered representative enough. Even assuming such a simple shape, the roughness alone of the ice surface makes the problem a complex one. Its theoretical effect on h has been studied in detail (Makkonen, 1985), and can be used as part of an icing model.

With an estimate of h, equation (4.14) can now be used to determine the accretion efficiency α_3, and thereby the rate of icing, equation (4.1). It should be noted that although equation (4.14) is set out in terms of the spray flux density F, it is basically valid also locally on the surface of an iced object. In such a case F is the direct mass flux plus the run-back water from the other sectors of the surface. Then also the mean temperature of the net flux will differ from that of the droplets. In order to predict not only the overall mass of the accretion but also its shape and vertical distribution, these aspects of formulating the local heat balance have been included in some recent icing models (e.g. Lozowski *et al.*, 1983; Szilder *et al.*, 1987).

4.2 MODELLING OF ICING

4.2.1 Climatological methods

In construction projects for masts, power lines etc., design ice loads are often determined by rough estimates of the severity of icing based on climatological data. Factors such as typical weather patterns, temperature regimes, cloud heights, elevations, types of terrain and distances from the coast can be used as indications of icing risk.

However, the prediction of ice loads from climatology entails some major difficulties. First, the basic data on ice loads are usually very limited. Second, interpolating and extrapolating such data is a difficult task even for an experienced meteorological or icing specialist, owing to the numerous influencing parameters and combined effects of different types of icing. For these reasons it is preferable to combine climatological analysis with numerical modelling (section 4.2.2) in such a way that the meteorological input parameters, and not the ice load, are interpolated or extrapolated for the construction site by weather stations in the vicinity. The ice loads can then be calculated for the site through numerical modelling or empirical equations (Haldar *et al.*, 1988; Makkonen and Ahti, 1994).

Unfortunately, the liquid water content and drop size necessary for the modelling of rime icing are not routinely measured at weather stations.

Estimation of these parameters for construction sites must be based mainly on the height of the cloud base (Ahti and Makkonen, 1982; Makkonen and Ahti, 1994) and on a subjective evaluation. Development of new mesoscale models of the atmospheric boundary layer will hopefully improve this situation in future.

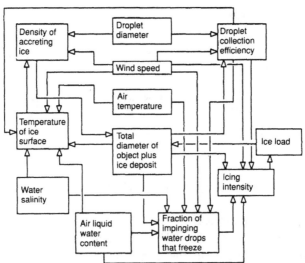

Fig. 4.6 Factorial interdependence in the icing process caused by water droplets.

4.2.2 Numerical modelling

Solving the icing rate analytically using equation (4.14) is impractical because this involves empirical equations for the dependence of saturation water pressure and specific heats on temperature, and the whole procedure for determining h. Numerical methods must be used, because icing is a time-dependent process, and changes in the dimensions of accretion affect, for example, the heat transfer coefficient. All this makes the process of icing determination rather complex. Figure 4.6 is a schematic representation for the many relationships involved.

Modern computers greatly facilitate the achievement and interpretation of results obtained from complex icing models. The problem of accretion shape changing with time is usually avoided by assuming that the ice deposit maintains its cylindrical geometry. The growth of icicles, however, may complicate the problem. Thus a separate model simulating icicle growth (Makkonen, 1988; Makkonen and Fujii, 1993) can be included when modelling icing caused by freezing rain. Stochastic aspects of the water flow and icing may also be included in the modelling (Szilder, 1994).

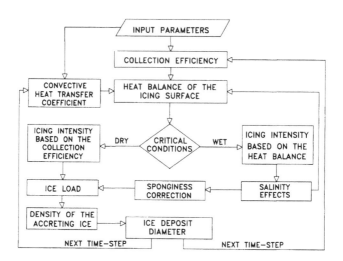

Fig. 4.7 Simplified block diagram of the Makkonen model (Makkonen, 1984b).

Time-dependent numerical models also require modelling of the density of accreted ice (section 4.3). This is because the icing rate for the next time-step is dependent on the dimensions of the object, A in equation (4.1), and the relationship between the modelled ice load and the dimensions of the structure is therefore required.

When estimates of the density of accretions are included in the system, a numerical model can be developed to simulate the time-dependent icing of an object. A schematic description of an icing model is shown in Fig. 4.7.

A real structure such as a mast usually comprises small structural members of different sizes. Modelling the icing of such a complex structure may be attempted by breaking the structure down into an ensemble of smaller elements, calculating the ice load separately for each element, and summing the results to obtain the total ice load. However, the validity of such modelling is questionable because of the shadowing effects of, and interactions between, different parts of the structure. Thus for complex structures physical modelling is preferable (section 4.2.3).

4.2.3 Physical modelling

When modelling the icing of complex structures, it is important to remember that some components of the structure may shelter others from ice ac

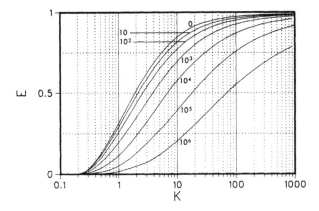

Fig. 4.8 Smooth curves drawn for the numerical solution of the total collision efficiency E as a function of the nondimensional parameters K and F (from Finstad *et al.* 1988a).

cretion. Also, different parts of the structure may freeze completely together and should thereafter be modelled as a single object. These aspects should be considered individually for each structure and can be studied by small-scale icing experiments (Makkonen and Oleskiw, 1994).

Langmuir and Blodgett (1946) have shown that droplet trajectories and the rime icing process can be described by two dimensionless parameters: the Stokes number K and the Reynolds number Re_d. These two parameters are given by

$$K = \frac{\lambda_w V d^2}{9\mu L} \tag{4.15}$$

where λ_w is the water density, d is the droplet diameter, V is the free stream velocity, μ is the dynamic viscosity of air, and L is the cross-sectional width of the object; and by

$$Re_d = \frac{V d \lambda_a}{\mu} \tag{4.16}$$

where λ_a is the density of air.

A convenient way to utilize this result of the dimensional analysis is to calculate the local and total droplet collision efficiencies, the droplet impact angle and the dimensionless droplet impact velocity V_0 (= v_0/V, where v_0 is the impact velocity) from the two dimensionless parameters K and Φ where $\Phi = Re_d^2/K$. The most recent methods used for this purpose are given by Finstad *et al.* (1988a). Their numerical result for the total collision efficiency is shown in Fig. 4.8.

A result of this dimensional analysis is that rime icing can be perfectly simulated, on any scale, providing that the parameters K and Re_d are kept constant. It follows from equations (4.15) and (4.16) that the similarity requirement for the modelling in terms of the droplet size and wind speed is

$$d_{\text{small-scale}} = \frac{d_{\text{full-scale}}}{\lambda}$$

$$V_{\text{small-scale}} = V_{\text{full-scale}} \cdot \lambda \tag{4.17}$$

where $\lambda = L_{\text{full-scale}}/L_{\text{small-scale}}$ is the geometric scale factor.

In long-term icing, the similarity of K and Re_d is not sufficient, because ice shape will not remain correct unless the density of the rime is the same at small scale and full scale. An approximate similarity factor for rime density is the Macklin parameter R (see equation (4.18) below).

When scaling down the icing process by keeping K and Re_d constant, the relationship $V_0 = v_0/V$ remains constant also: that is, v_0 will scale as V and R as $V \cdot d_m$. Considering that the collision efficiency can be determined quite accurately using the median volume diameter d_m of the droplet spectrum (Finstad *et al.*, 1988b), Re_d also scales as $V \cdot d_m$. Thus by correctly scaling R, the density of rime will remain constant provided that t_s in equation (4.19) remains constant as well. The surface temperature t_s can easily be kept constant in the scaling because it is dependent upon the air temperature and water flux, both of which are effectively independent of K and Re_d (Makkonen, 1984b).

A conclusion of the above discussion is that, in theory, rime icing can be perfectly simulated at a small scale. Even studying long-term icing of complex structures in detail is therefore theoretically possible with small-scale experiments.

Unfortunately, it is extremely difficult to meet the theoretical similarity criteria in practical small-scale experiments. For example, simulating rather typical rime icing conditions:

$d_{\text{m full-scale}} = 12 \ \mu\text{m}$

$V_{\text{full-scale}} = 20 \ \text{m s}^{-1}$

while using a scale factor of $\lambda = 15.3$ (Makkonen and Oleskiw, 1994) would require, according to equation (4.17), the use of

$d_{m \cdot small\text{-}scale} = 1.3 \ \mu m$ and

$V_{small\text{-}scale} = 300 \ m \ s^{-1}$

This velocity is so high that any rime accreted would be immediately eroded or blown away.

Despite the above limitation, the theoretical scaling laws can be utilized as guidelines while attempting to minimize the collision efficiency (Makkonen and Oleskiw, 1993). The theory suggests that when it is difficult to reduce E via reductions in d_m, it is more effective to reduce K than Φ. Based upon this, a low wind speed can be chosen. Under the circumstances, an evaluation of the success of the scaling validity must be based upon comparisons between the resulting overall small-scale icing rates and ice shapes with those found full-scale in nature.

In summary, perfect simulation of rime icing (dry growth) is possible in theory, but not in practice. However, very good approximate scaling of icing can be done, so that valuable information on the icing process on large structures can be obtained. The technique is particularly useful in studying the effect of structure design and orientation on rime ice loads. Physical modelling of glaze icing (wet growth) has not been attempted, the main difficulty being the high wind speed required to scale the heat transfer coefficient properly (section 4.3.2).

4.3 PROPERTIES OF ACCRETED ICE

From a practical point of view, the most important property of accreted ice is its density. This is because, in the modelling of icing, the icing rate for the next time-step depends on the dimensions of the object, and the relationship between the modelled ice load and the dimensions of the iced structure is, therefore, required.

The following best-fit equation (Makkonen and Stallabrass, 1984) is recommended for the density ρ of rime ice (dry growth):

$$\rho = 0.378 + 0.423(\log R) - 0.0823(\log R)^2 \tag{4.18}$$

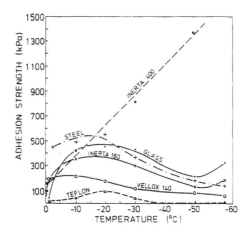

Fig. 4.9 Dependence of adhesion strength of bulk ice on air temperature.

where R is Macklin's (1962) parameter:

$$R = -\left(\frac{v_0 d_\mathrm{m}}{2t_\mathrm{s}}\right) \qquad (4.19)$$

where v_0 is the droplet impact speed based on the median volume droplet size d_m, and t_s is the surface temperature of the accretion.

Equations for calculating v_0 can be found elsewhere (Finstad *et al.*, 1988a). The surface temperature t_s must be solved numerically from the heat balance equation. However, in most cases of atmospheric rime icing, t_s can be approximated by the air temperature t_a.

A qualitative result is obtained for equations (4.18) and (4.19) in that the density of rime ice rises rapidly with decreasing object size and increasing wind speed, drop size and air temperature.

For glaze ice (wet growth), the density variations are small and the value of 0.9 g cm^{-3} can be assumed: wet snow accretions typically have a density of 0.3–0.4 g m^{-3} (Koshenko and Bashirova, 1979).

The adhesion strength of ice is generally so high that spontaneous drop-off of even the thickest accreted ice layers from structural surfaces cannot be expected as long as the temperature remains below 0 °C. The dependence of the adhesion strength of bulk ice on temperature for various materials is shown in Fig. 4.9.

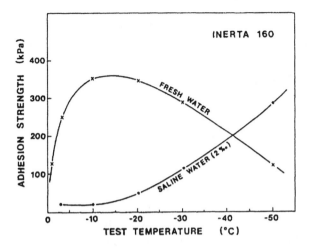

Fig. 4.10 Dependence of adhesion strength on air temperature in marine conditions.

Wind tunnel studies on the adhesion of ice formed by droplet accretion (Laforte *et al.*, 1983) indicate that microscale adhesion is the same as for bulk ice. Thus differences in the adhesion strength of ice accreted under different atmospheric conditions originate from differences in the real contact area, which is smaller for low-density ice.

In marine conditions, the salt in spray water considerably affects the adhesion strength (Fig. 4.10). At higher temperatures the adhesion strength is much lower for saline ice, but at very cold temperatures is greater than for freshwater ice (Makkonen, 1989).

Other strength properties of accreted ice – such as compressive and shear strength – may also be of interest in de-icing, for example. Schaefer *et al.* (1989) have suggested that hardness testing by indenters gives the most relevant indication of these properties.

Crystal size close to the object surface (that is, in the initial stage of the icing process) is largely determined by the properties of the substrate (Mizuno, 1981). At some distance (5–29 mm) from the substrate, the effect of the substrate becomes minimal, and beyond that point the crystal structure is determined only by the growth conditions. In practice this means that the mean crystal size increases with distance above the surface. Crystal size decreases with falling air temperature t_a. Droplet size, impact speed, and liquid water content seem to have only a small effect on the crystal size (Rye and Macklin, 1975). However, Prodi *et al.* (1982) observed that the average surface area of the grains is smaller in wet-spongy accretions than in dry ones grown at the same air temperatures and at deposit temperatures just below 0 °C, which implies that the liquid water content has some effect on crystal size, at least near the critical conditions.

In the case of freezing rain, the droplets solidify into single crystals (monocrystallization). Polycrystallization is also possible, but at lower air temperatures (Hallett, 1964; Mizuno, 1981), in which case the number of crystals nucleated from one droplet increases with decreasing air temperature.

Crystals in wet snow deposits are smaller than in glaze. This is probably related to the dimensions of ice crystals in the original snowflakes.

The c-axis orientation of an ice crystal is close to perpendicular to the direction of growth direction in accretions grown in wet-growth conditions, and parallel to it if grown in dry-growth conditions (Levi and Aufdermaur, 1970). At very low air temperatures the c-axis orientation in dry growth tends to shift towards 45°.

4.4 DISCUSSION

The need to consider ice accretion arises when planning construction in mountainous areas and cold marine environments. The weight of ice on a lattice mast can be as much as 1.5 t per metre of mast, exceeding the weight of the structure itself. On power line cables, weights of more than 100 kg per metre have been observed under extreme icing conditions. Loads of up to 500 t have been estimated on offshore drilling rigs.

In addition to the load formed by accreted ice, there is that caused by the consequent increase in wind drag, especially on tall masts. The wind load can easily increase by a factor of 5 on a fully iced lattice structure. Engineering aspects of accretion of ice on structures have been discussed during a series of International Workshops on Atmospheric Icing of Structures.

The theory of ice accretion on structures has been partly verified (Makkonen and Stallabrass, 1984, 1987; Gates *et al.*, 1986), but several areas still require further development and verification.

One major uncertainty concerns the collision efficiency α_1. When this is very small (< 0.1), the theory (section 4.1.3(a)) tends to predict too small a value (Personne and Gayet, 1988), mainly because the roughness elements of the surface act as individual collectors. When α_1 is small the icing rate is also very low, so this problem does not generally hamper the estimation of design ice loads. However, when the size of the structure, *A* in equation (4.1), is large, such as for a fully iced mast, the growth rate of the total ice load may be substantial, even at low values of α_1. Thus estimates of icing for very large objects, especially at low wind speeds, should be made with caution.

Estimation of the sticking efficiency α_2 of wet snow is also inaccurate at present. Equation (4.6) should be seen merely as a first approximation until more sophisticated methods for estimating α_2 are developed. For large water drops (rain) there is still the possibility that some might bounce (List, 1977), in which case assuming $\alpha_2 = 1$ may be a source of error.

The accretion efficiency α_3 is generally the most accurate factor in equation (4.1). The theoretical estimation of glaze formation (wet growth) is relatively reliable, provided that the model has the correct input.

The theory is mostly based on the assumption that the shape of the icing object is cylindrical. In the field, structural members may not be cylindrical, and if they are they may not remain so after some time of ice accretion. This causes errors in numerical modelling. There are indications, however, that this is not a major problem when predicting ice loads (Makkonen, 1984b; Makkonen and Stallabrass, 1984), unless deviation from the cylindrical shape is extreme. Methods for predicting the shape of ice accretion have been developed (e.g. Lozowski *et al.*, 1983; Szilder *et al.*, 1987; Szilder, 1994), but they are of little use until factors α_1, α_2 and α_3 in equation (4.1) can be predicted for more complex shapes. The shape of the accretion is, however, important regarding the wind drag.

As to the use of theoretical icing models for predicting design ice loads, one major problem is the input requirement. The median volume droplet size (MVD) and liquid water content (LWC), which are not routinely measured, are of less significance when considering freezing precipitation icing (Makkonen, 1981), but critically affect rime icing. Extrapolation of these and other required input parameters to the often remote sites of the structures of interest is extremely difficult. The future usefulness of the theoretical modelling of icing depends essentially on progress in this area. This also applies to offshore spray icing, where methods for estimating the formation and amount of sea-spray ice are still unsatisfactory.

REFERENCES

Admirat, P., Fily, M. and Goncourt, B. de, (1986) *Calibration of a Wet Snow Model with 13 Natural Cases from Japan.* Technical Note, Electricité de France, Service National Electrique, 59 pp.

Ahti, K. and Makkonen, L. (1982) Observation on rime formation in relation to routinely measured meteorological parameters. *Geophysica,* **19**, 75–85.

Finstad, K.J., Lozowski, E.P. and Gates E.M. (1988a) A computational investigation of water droplet trajectories. *Journal of Atmospheric Oceanic Technology*, **5**, 160–70.

Finstad, K.J., Lozowski, E.P. and Makkonen, L. (1988b) On the median volume diameter approximation for droplet collision efficiency. *Journal of Atmospheric Science*, **45**, 4008–12.

Gates, E.M., Narten, R., Lozowski, E.P. and Makkonen, L. (1986) Marine icing and spongy ice, in *Proceedings of 8th International Symposium on Ice, International Association for Hydraulic Research (IAHR),* Iowa City, USA, Vol. II, pp. 153–63.

Haldar, A., Mitten, P. and Makkonen L. (1988) Evaluation of probabilistic climate loadings on existing 230 kV steel transmission lines, in *Proceedings of 4th International Conference on Atmospheric Icing of Structures*, 5–7 September, 1988, Paris, pp. 19–23.

Hallett, J. (1964) Experimental studies of the crystallization of supercooled water. *Journal of Atmospheric Science*, **21**, 671–82.

Koshenko, A.M. and Bashirova, L. (1979) Recommendations on forecasting the precipitation and deposition (sticking) of wet snow. *Trudy UkrNiGMI*, **176**, 96–102 (in Russian).

Laforte, J.-L., Phan, C.L., Felin, B. and Martin, R. (1983) Adhesion of ice on aluminium conductor and crystal size in the surface layer, in *Proceedings of First International Workshop on Atmospheric Icing of Structures* (ed. L.D. Minsk). US Army Cold Regions Research and Engineering Laboratory, Special Report 83-17, pp. 83–92.

Langmuir, I. and Blodgett, K.B. (1946) *A Mathematical Investigation of Water Droplet Trajectories.* Technical Report 54118, USAAF, 65 pp.

Levi, L. and Aufdermauer, A.N. (1970) Crystallographic orientation and crystal size in cylindrical accretions of ice. *Journal of Atmospheric Science*, **27**, 443–52.

List, R. (1977) Ice accretion on structures. *Journal of Glaciology*, **19**, 451–65.

Lozowski, E.P., Stallabrass, J.R. and Hearty, P.F. (1983) The icing of an unheated, nonrotating cylinder. Part I: A simulation model. *Journal of Climate and Applied Meteorology*, **22**, 2053–62.

Macklin, W.C. (1962) The density and structure of ice formed by accretion. *Quarterly Journal of Royal Meteorological Society*, **88**, 30–50.

Makkonen, L. (1981) Estimating intensity of atmospheric ice accretion on stationary structures. *Journal of Applied Meteorology*, **20**, 595–600.

Makkonen, L. (1984a) *Atmospheric Icing on Sea Structures*. US Army Cold Regions Research and Engineering Laboratory Monograph 84–2, 102 pp.

Makkonen, L. (1984b) Modeling of ice accretion on wires. *Journal of Climate and Applied Meteorology*, **23**, 929–39.

Makkonen, L. (1985) Heat transfer and icing of a rough cylinder. *Cold Regions Science and Technology*, **10**, 105–16.

Makkonen, L. (1987) Salinity and growth rate of ice formed by sea spray. *Cold Regions Science and Technology*, **14**, 163–71.

Makkonen, L. (1988a) A model of icicle growth. *Journal of Glaciology*, **34**. 64–70.

Makkonen, L. (1988b) Estimation of wet snow accretion on structures. *Cold Regions Science and Technology*, **17**, 83–8.

Makkonen, L. (1989) Adhesion of pure ice and saline ice at temperatures −1 to −60 °C, in *12th Annual Meeting of the Adhesion Society*, Hilton Head Island, SC, USA, 20–22 February 1989, pp. 23a–23d.

Makkonen, L. (1990) The origin of spongy ice, in *10th International Association for Hydraulic Research (IAHR) Symposium on Ice*, Helsinki, Finland, Vol. II, pp. 1022–30.

Makkonen, L. and Ahti, K. (1994) Climatic mapping of ice loads based on airport weather observations. *Atmospheric Research* (in press).

Makkonen, L. and Fujii, Y. (1993) Spacing of icicles. *Cold Regions Science and Technology*, **21**, 317–22.

Makkonen, L. and Oleskiw, M. (1994) Small-scale experiments on rime icing. *Cold Regions Science and Technology* (in press).

Makkonen, L. and Stallabrass, J.R. (1984) *Ice accretion on cylinders and wires*. National Research Council of Canada, NRC, Technical Report. TR-LT-005, 50 pp.

Makkonen, L. and Stallabrass, J.R. (1987) Experiments on the cloud droplet collision efficiency of cylinders. *Journal of Climate and Applied Meteorology*, **26**, 1406–11.

Mizuno, Y. (1981) Structure and orientation of frozen droplets on ice surfaces. *Low Temperature Science, Series A*, **40**, 11–23 (in Japanese).

Personne, P. and Gayet, J.-F. (1988) Ice accretion on wires and anti-icing induced by Joule effect. *Journal of Climate and Applied Meteorology*, **27**, 101–14.

Prodi, F., Levi, F., Frantini, A. and Scarani, C. (1982) Crystal size and orientation in ice grown by droplet accretion in wet and spongy regimes. *Journal of Atmospheric Science*, **39**, 2301–12.

Rye, P.J. and Macklin, W.C. (1975) Crystal size in accreted ice. *Quarterly Journal of Royal Meteorological Society*, **101**, 207–15.

Schaefer, J.A., Ettema, R. and Nixon, W.A. (1989) Measurement of icing hardness. *Cold Regions Science and Technology*, **17**, 89–93.

Szilder, K. (1994) Simulation of ice accretion on a cylinder due to freezing rain. *Journal of Glaciology* (in press).

Szilder, K., Lozowski, E.P. and Gates, E.M. (1987) Modelling ice accretion on non-rotating cylinders – the incorporation of time dependence and internal heat conduction. *Cold Regions Science and Technology*, **13**, 177–91.

5

Abrasion of concrete structures by ice

5.1 ABRASION MECHANISMS

A concrete offshore structure in arctic conditions is subjected to various damage and load effects. On the basis of their effect, they can be classified as mechanical, physical or chemical action which causes damage to concrete. The damage effects are set out in Fig. 5.1.

When a moving ice sheet breaks against a structure it causes abrasion in the concrete. If the concrete aggregate particles are protruding, forces have various directions, depending on the route of the ice in relation to the concrete structure surface. Ice frozen onto the concrete may, in addition, cause external mechanical forces in the aggregate. The magnitudes of the forces depend both on the ice properties and the size of aggregate particles. Loads due to ice occur depending on the ice failure mode. Physical damage is attributable to the pressure of freezing water present in the concrete and the shrinkage of concrete, as well as to thermal gradients. Shrinkage as well as temperature gradients cause cracks in the concrete, which permit the penetration of moisture and salts. Naturally, shrinkage does not occur if the structure is continuously in contact with water after hardening of the concrete.

In this text the abrasion problem is treated mainly in sea conditions of Nordic countries. The subject has been studied also in Japan (Itoh *et al.*, 1989, 1994), in Canada, and in the USA (Nawwar and Malhotra, 1988; Hoff, 1989).

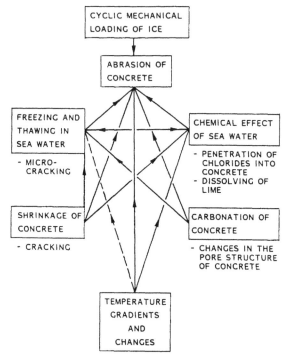

Fig. 5.1 Effects causing damage to concrete in seawater.

5.2 ABRASION TESTING

5.2.1 General

The solution to the abrasion problem of concrete sea structures has been sought both by developing abrasion-testing methods for laboratory use and by measuring the abrasion depths of lighthouses and other sea structures. The friction forces between concrete and ice have also been measured. The most dominant factors found to affect abrasion are the temperature of the ice, the stress intensity of the ice against the concrete surface in a friction test, the strength of concrete and, especially, the durability of the strength during freeze–thaw cycles. The lower the temperature of the ice and the higher the stress intensity, the greater is the abrasion. The ice abrasion problem, however, is more complex than a simple abrasion of the concrete measured in a few tests (Janson, 1988; Huovinen, 1990).

In a Finnish study (Huovinen, 1990) the determination of abrasion of concrete in arctic offshore structures was based on four different methods:

1. laboratory tests;
2. tests with an icebreaker;
3. abrasion studies on Finnish lighthouses;
4. computer calculations.

An abrasion machine was developed for laboratory use. The abrasion resistance of different concretes can be studied with the abrasion machine so that the concrete will have under gone cyclic freezing–thawing tests before the abrasion tests.

The abrasion resistance of similar concrete mixes was studied at sea with an icebreaker. In icebreaker tests the specimens were fastened onto the bow of the icebreaker at water level. The abrasion of the concrete specimens was measured at the end of the tests. The abrasion of Finnish lighthouses was measured at four lighthouses.

The abrasion and fracture of the concrete were also studied with computer calculations. The ice pressures against small areas, such as aggregate particles protruding from the surface of a concrete structure, were measured with laboratory tests. Also, the bond strength between aggregate particles and hardened cement paste was measured in tests. These values were needed in the computer calculations. On the basis of the calculations, using the calculation model, the abrasion of concrete was estimated as a function of ice sheet movement (Huovinen, 1990).

5.2.2 Field studies

The field investigation showed that ice had abraded the concrete significantly. The abrasion of the lighthouse in Helsinki was measured to be about 300 mm over 30 years. The lighthouses in the Gulf of Bothnia had abrasions of between 22 and 24 years. The principal reasons for the abrasion were the low resistance of concretes to frost in combination with the abrasion by ice. The ice had worn off the part of the concrete surface initially damaged by frost. Subsequently, the new surface had been damaged during the repeated freezing and thawing. In this way the damage mechanism is repeated.

The compressive strength of the lighthouse concretes was measured at the surface of the structure, both at water level and at 1.5 m above water level. The strengths of concrete at water level were found to be 53–58% of the strengths of concrete above water level.

In a Swedish abrasion study of concrete lighthouses in the Baltic Sea, the abrasion problem was studied both by measuring the abrasion depths in, and analysing the concrete cores of, the lighthouses. The following formula shows the relationship between abrasion rate and ice drift velocity, ice thickness and time (Janson, 1988):

$$\text{Abrasion rate, ABR / year} = v_i h_i dt + 0.08 \quad (\text{mm / year}) \tag{5.1}$$

where v_i is the ice drift velocity in knots, h_i is the ice thickness in mm, and t is the time in days.

According to the results of the field study, the abrasion rate varied from 0.2 to 7.0 mm per year. The abrasion rate is obtained by dividing the maximum abrasion depth by the number of years.

The abrasion of concretes was also studied at sea with an icebreaker. The test specimen was placed on the bow of the icebreaker at water level. The test arrangements are presented in Fig. 5.2. The abrasion of the concrete on the surface of the specimens varied between 2 and 15 mm (mean values) when the compression strength varied between 30 and 60 MPa and the movement of the icebreaker was about 40 km.

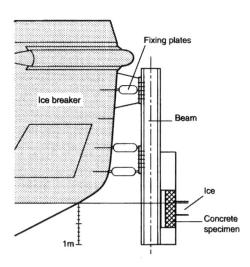

Fig. 5.2 Test arrangements of icebeaker test.

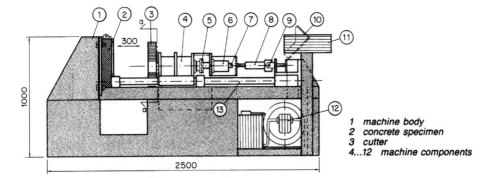

Fig. 5.3 Abrasion machine and test specimen.

5.2.3 Laboratory studies

(a) Abrasion durability
In the Finnish study an abrasion machine was developed for laboratory use, with the help of which the abrasion resistance of concrete was examined after a varying number of freeze–thaw cycles. Figure 5.3 shows the structure of the abrasion machine and the slab subjected to abrasion.

(b) Concrete strength versus freeze–thaw cycles
The changes of the strength and strain properties of different concretes during freeze–thaw cycles were examined by means of laboratory tests. The test specimens were frozen and thawed in synthetic seawater corresponding to ocean water. The temperature varied from −40 °C to +20 °C. The concretes were subjected to 50–100 freeze–thaw cycles. The properties examined comprised concrete strengths, such as compressive and tensile strength, and the bond strength between the aggregate and the cement paste, which is critical in abrasion of concrete when ice presses against the structure and breaks.

Other factors studied in the Finnish work were the stress–strain relationships as well as the changes of fracture energy measured in the compression test during the freeze–thaw cycles. In addition, the reduction in the fatigue strength of concretes due to repeated freezing and thawing was studied by means of mechanical cyclic loading tests. The studied concrete mixtures were

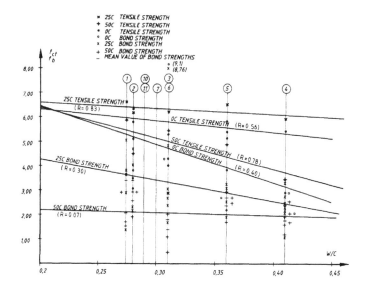

Fig. 5.4 Flexural tensile strength of concrete and the bond strength of aggregate particles as a function of the water/cement ratio and number of freeze–thaw cycles.

1. ordinary cement concrete (binder amount 500 kg m^{-3});
2. blast-furnace cement concrete (binder amount 500 kg m^{-3});
3. Portland cement concrete with added silica fume (binder amount 420 kg m^{-3} and about 9% silica fume);
4. concrete with lightweight aggregates (binder amount 438 kg m^{-3}).

The entrained air in the concrete was varied (3–8%). The design strength of the concretes was 60 MPa apart from the concrete made with lightweight aggregates (30 MPa).

The best results in the strength tests after freeze–thaw cycles were achieved with concretes containing silica fume and blast-furnace slag and the worst results were with lightweight aggregate concretes. The fracture energies decreased in all the concretes during the freeze–thaw cycles.

In the case of ordinary cement concretes, the bond strength of aggregate particles at the concrete surface is reduced during the repeated freezing–thawing tests more rapidly than the compressive or tensile strengths. The ratio of the bond and tensile strengths was about 0.7 at the beginning of the freezing–thawing tests and about 0.3–0.5 at the end of the tests.

In Fig. 5.4 the flexural tensile strength of test concretes and the bond strength of aggregate particles are presented as functions of water/cement ratio after 25 and 50 freeze–thaw cycles. Regression lines have also been calculated for the test results.

(c) Ice forces against protruding aggregate particles

In the Finnish study the ice forces against protruding aggregate particles were also measured with laboratory tests. Both the shear component parallel to the concrete surface and the normal component perpendicular to the surface were measured. Figure 5.5 shows the force components. The values are needed in the abrasion calculations. The values of the loads (impact speed 0.5 m s⁻¹) are presented in Table 5.1 for ice sheet thickness $d = 100$–1500 mm when the air temperature is -40 °C and seawater temperature -2 °C.

Table 5.1 Loads due to ice impact on protruding aggregate particles

Stone size (mm dia)	Normal component, σ	Shear component, Φ
8	17	17
32	10	17

5.3 MODELLING OF ABRASION DEPTH

The calculation models for the abrasion of concrete are presented in Fig. 5.6. The elements of the calculation models are parametric elements with eight nodes. The crack length L_{cr} is calculated by increasing the loads little by little to the values corresponding to σ_1 and Φ_1. The effect of ice is inserted in the calculation model to act as external load. The loads are produced by a normal component perpendicular to the concrete surface and

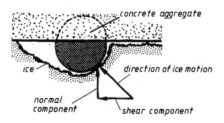

Fig. 5.5 Forces against protruding aggregate particles.

also by a shear component parallel to the surface when fine binder and aggregate particles have worn off and large aggregate particles protrude from the surface. The magnitudes of these force components can be determined in laboratory tests (Huovinen, 1990). The material constants and the strength and strain values are determined in laboratory tests and they were also inserted into the model.

The calculation model (a) (protrusion 0.7R) is chosen on the basis of studies of the surfaces of actual concrete offshore structures. On the basis of the studies the protrusion of the aggregate particles on the surface of the offshore structure at water level is generally 0.7R when R is the radius of the particle. The calculation model (b) (protrusion R) is chosen to cover the loosening of the stone from the surface.

The following formula takes into account the recurrence of ice loads in damp concrete (Cornelissen, 1984).

$$\log N = 13.92 - 14.42 \frac{\sigma_{max}}{f_{ab}} \qquad (5.2)$$

where N is the number of impacts by ice, σ_{max} is the tensile strength of the transition layer between the aggregate particle and cement paste subjected to repeated loading, and f_{ab} is the tensile strength of the transition layer between the aggregate particle and cement paste subjected to static load.

In the computer calculations the following values have been used:

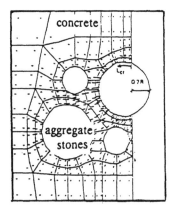 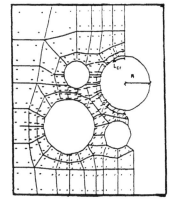

Fig. 5.6 Calculation models for the abrasion of concrete.

E_c = 5 000 $\sqrt{f_c}$ for concrete (modulus of elasticity for concrete measured
   in test);
E_{ctr} = 30 000 MPa (modulus of elasticity for transition layer);
E_a = 50 000 MPa (modulus of elasticity for aggregate);
ε_{fc} = 2.2‰ for f_c (yield strain of concrete);
ε_{cu} = 3.5‰ (ultimate strain of concrete);
f_a = 200 MPa (aggregate strength in compression);
f_{at} = 14 MPa (aggregate strength in tension);
f_{ct} = 0.1f_c (concrete strength in tension);
f_{ab} = 0.9f_{ct} (bond strength between aggregate particles and cement
   paste).

The abrasion depth of concrete in arctic sea structures can be calculated
as the sum total of abrasion depth of cement paste measured in icebreaker
tests at sea and the loosening of aggregate particles from surface
(Huovinen, 1990).
   The condition for the loosening of the aggregate particle can be considered as

$$\frac{L_{cr}}{R} = 1 \qquad\qquad (5.3)$$

where R is the radius of the aggregate stone, and L_{cr} is the crack length.
   The abrasion rate for hardened cement paste is achieved with the results
in the icebreaker test:

$$b = \frac{3}{f_c}s \ (\text{mm km}^{-1}) \qquad\qquad (5.4)$$

where s is the movement of the ice sheet (km), and f_c is the compressive
strength of concrete (MPa).
   The total abrasion depth can be calculated with the formula

$$\text{ABR} = \sum_{i=1}^{n} a_i \frac{\log n_s}{\log n_1} R_i + \left(1 - \sum a_i\right) \cdot b \qquad\qquad (5.5)$$

where a_i is the proportional amount of aggregate particles of radius R_i, n_s is
the number of ice impacts during ice sheet movements, n_1 is the number of

ice impacts when $L_{cr}/R = 1$, and b is the abrasion rate of hardened cement paste (mm).

The abrasion diagrams are valid when the aggregate distribution of concrete is normal ($\phi \leq 6$ mm 43%, $6 \leq \phi \leq 12$ 18%, $12 \leq \phi \leq 24$ 26%, $24 \leq \phi \leq 32$ 13%). In addition to the compressive strength requirement $f_c = 40$, 60, 80, 100 MPa, it is presupposed that the tensile strength of concrete f_{ct} is at least 10% of the compressive strength, and the bond strength between aggregate particles is at least 90% of the tensile strength of concrete.

The abrasion depths calculated with equation (5.5) for compressive strengths $f_c = 40$, 60, 80 and 100 MPa are presented in Fig. 5.7.

5.4 COMPARISON OF TEST RESULTS AND COMPUTER CALCULATIONS

In Fig. 5.8 a comparison of the abrasion (maximum and minimum values)

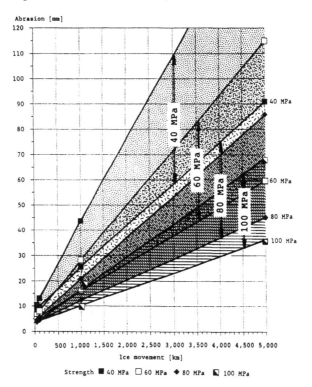

Fig. 5.7 Abrasion of concrete strength $f_c = 40$, 60, 80 and 100 MPa as a function of ice sheet movement.

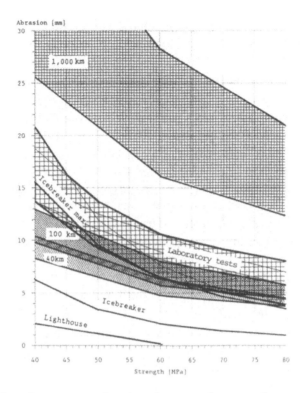

Fig. 5.8 Abrasion of concrete as a function of compressive strength.

of concrete is presented as a funcion of the compressive strength of concrete in laboratory abrasion tests run for 10 min, in icebreaker tests, for ice
field movements of 40, 100 and 1000 km according to the abrasion calculations and in the Finnish lighthouses for one year.

5.5 CONCLUSIONS

The following conclusions can be drawn:

1. The abrasion depth and resistance of concrete in arctic sea conditions
   can be determined either by using abrasion diagrams or by laboratory
   tests. The best results are achieved by using both these methods.
   When using the abrasion diagrams, the abrasion depth can be
   estimated as a function of the compressive strength of concrete and ice
   sheet movement. The abrasion resistance can also be measured by
   using the laboratory abrasion test lasting 10 min. Before the abrasion

test, the concrete plate should undergo a cyclic freeze–thaw test in seawater for 50 cycles with temperature varying between −50 and +20°C.

2. The most important mechanical factor pertaining to the measurement of the resistance to abrasion of concrete is the strength of the concrete. The bond strength of aggregate particles and cement paste as well as its resistance to repeated freeze–thaw cycles is especially crucial. For a good abrasion resistance the compressive strength of concrete should be at least $f_c = 70$ MPa. In addition, the concrete must naturally be frost-resistant. In the tests performed, the strength of concrete during the repeated freeze–thaw tests was best when the maximum water/cement ratio of the concretes was 0.3–0.35.

3. In the case of ordinary cement concretes, the bond strength of aggregate particles at the concrete surface decreases during repeated freeze–thaw tests more rapidly than the compressive or tensile strengths. The relation between the bond and tensile strengths was about 0.7 at the beginning of the freeze–thaw tests and about 0.3–0.5 at the end of the tests in ordinary cement concretes. Apparently, the changes in temperature, when they exceed approximately $\Delta T = 40°C$, deteriorate the bond of the particles at the surface and increase cracking, especially in the bond zone of the aggregate particles.

4. The best results in both the strength and abrasion tests are achieved with concretes containing silica fume and blast-furnace slag.

5. The resistance to abrasion of concrete can be improved by preventing frost damage by keeping the entire wall either so warm or so frozen that it is not exposed to freeze–thaw cycles.

6. If the value obtained for the bond strength of concrete aggregate particles is at least $f_{ab} = 8$ MPa, the resistance to ice abrasion of concrete is considered very good.

7. If the abrasion depth in the laboratory test after a freeze–thaw test of 50 cycles is at the most 10 mm, the resistance to ice abrasion of concrete can be considered good.

8. Increasing the maximum size of the aggregate reduces the abrasion in concrete. Large particles protruding from the concrete surface break the ice before the ice affects the finer components of the concrete.

9. The use of hard homogeneous concrete in the ice abrasion zone reduces abrasion because the surface is subjected to even abrasion and there are no detaching aggregate particles.

REFERENCES

Cornelissen, H. (1984) Fatigue failure of concrete in tension. *Heron*, Delft University of Technology, Netherlands, **29** (4).

Hoff, G.C. (1989) Evaluation of ice abrasion of high-strength lightweight concretes for arctic application, in *Proceedings of 8th International Conference on Offshore Mechanical and Arctic Engineering*, The Hague, pp. 583–590.

Huovinen S. (1990) *Abrasion of Concrete by Ice in Arctic Sea Structures.* Publication No. 62, Technical Research Centre of Finland.

Itoh, Y., Asai, Y. and Saeki, H. (1989) An experimental study of abrasion of various concrete due to sea ice movement, in *Proceedings of Evaluation of Materials Performance in Severe Environments*, EVALMAT 89, pp. 549–590.

Itoh, Y., Tanaka, Y. and Saeki, H. (1994) Estimation method for abrasion of concrete structures due to sea ice movement, in *Proceedings of Fourth International Offshore and Polar Engineering Conference*, Vol. II, pp. 545–552.

Janson, E. (1988) Long term resistance of concrete offshore structures in ice environment, in *7th International Conference on Offshore Mechanics and Arctic Engineering*. Houston, Texas, 7–12 February, American Society of Mechanical Engineers. Vol. III. pp. 225–30.

Nawwar, A.M. and Malhotra, V.M. (1988) Development of a test method to determine the resistance of concrete to ice abrasion and/or impact, Publication SP-109, American Concrete Institute, pp. 401–426.

6

Concluding summary and future needs

Ice is a peculiar material, as discussed in this report, particularly in section 2.2. There is still clearly insufficient understanding of ice as a material. This state of affairs requires further theoretical and experimental research also at basic level.

Regardless of the many poorly understood aspects of ice, it is the opinion of RILEM Technical Committee TC-118 that the potential for using ice as a construction material is presently underestimated. The only widely used application is construction of ice islands, while many other promising areas exist where ice could be effectively used. Ice is inexpensive, quick to make and can be protected from melting. Making ice in large amounts seems to be economically feasible only by utilizing a natural cold environment. This of course limits the effective use of ice in construction to locations where sub-freezing temperatures prevail for a considerable part of the year.

The strength and fracture toughness of ice can be increased by the addition of appropriate reinforcing materials. These reinforcers can be either large-scale (like steel rods) and placed in specific locations within the ice, or small-scale (sawdust and wood fibres) and dispersed evenly throughout the ice. There are a number of problems associated with the use of reinforcement in ice, but the strength benefits that they offer would make them ideal in certain situations. Further work is needed before design guides for the use of reinforced ice can be developed.

Bearing capacity, along with other ice properties, still requires further research. In particular, composites of ice and other materials should be studied in more detail.

The problems caused by ice in the form of accreting ice on structures and abrasion of concrete by moving ice sheets are severe. However, understanding of these phenomena has improved significantly during the last few years. Numerical modelling provides an effective means to analyse these problems for optimum design of structures in cold areas.

Index

Page numbers in **bold** type indicate illustrations; numbers in *italic* type indicate tables.

RILEM

RILEM, The International Union of Testing and Research Laboratories for Materials and Structures, is an international, non-governmental technical association whose vocation is to contribute to progress in the construction sciences, techniques and industries, essentially by means of the communication it fosters between research and practice. RILEM activity therefore aims at developing the knowledge of properties of materials and performance of structures, at defining the means for their assessment in laboratory and service conditions and at unifying measurement and testing methods used with this objective.

RILEM was founded in 1947, and has a membership of over 900 in some 80 countries. It forms an institutional framework for cooperation by experts to:

- optimise and harmonise test methods for measuring properties and performance of building and civil engineering materials and structures under laboratory and service environments;
- prepare technical recommendations for testing methods;
- prepare state-of-the-art reports to identify further research needs.

RILEM members include the leading building research and testing laboratories around the world, industrial research, manufacturing and contracting interests as well as a significant number of individual members, from industry and universities. RILEM's focus is on construction materials and their use in buildings and civil engineering structures, covering all phases of the building process from manufacture to use and recycling of materials.

RILEM meets these objectives though the work of its technical committees. Symposia, workshops and seminars are organised to facilitate the exchange of information and dissemination of knowledge. RILEM's primary output are technical recommendations. RILEM also publishes the journal *Materials and Structures* which provides a further avenue for reporting the work of its committees. Details are given below. Many other publications, in the form of reports, monographs, symposia and workshop proceedings, are produced.

Details of RILEM membership may be obtained from RILEM, École Normale Supérieure, Pavillon du Crous, 61, avenue du Pdt Wilson, 94235 Cachan Cedex, France.

RILEM Reports, Proceedings and other publications are listed below. Full details may be obtained from E & F N Spon, 2-6 Boundary Row, London SE1 8HN, Tel: (0)71-865 0066, Fax: (0)71-522 9623.

Materials and Structures

RILEM's journal, *Materials and Structures*, is published by E & F N Spon on behalf of RILEM. The journal was founded in 1968, and is a leading journal of record for current research in the properties and performance of building materials and structures, standardization of test methods, and the application of research results to the structural use of materials in building and civil engineering applications.

The papers are selected by an international Editorial Committee to conform with the highest research standards. As well as submitted papers from research and industry, the Journal publishes Reports and Recommendations prepared buy RILEM Technical Committees, together with news of other RILEM activities.

Materials and Structures is published ten times a year (ISSN 0025-5432) and sample copy requests and subscription enquiries should be sent to: E & F N Spon, 2-6 Boundary Row, London SE1 8HN, Tel: (0)71-865 0066, Fax: (0)71-522 9623; or Journals Promotion Department, Chapman & Hall Inc, One Penn Plaza, 41st Floor, New York, NY 10119, USA, Tel: (212) 564 1060, Fax: (212) 564 1505.

RILEM Reports

RILEM Proceedings

RILEM Recommendations and Recommended Practice

Milton Keynes UK
Ingram Content Group UK Ltd.
UKHW040050071024
449327UK00019B/458

9 780367 449285